Operations Excellence Management System (OEMS)

Operations Excellence Management System (OEMS)

Getting It Right the First Time

Authored by Chitram Lutchman,
Kevan Lutchman, Ramakrishna Akula,
Charles Lyons, and Waddah S. Ghanem Al Hashmi

CRC Press
Taylor & Francis Group
Boca Raton London New York

CRC Press is an imprint of the
Taylor & Francis Group, an **informa** business

CRC Press
Taylor & Francis Group
6000 Broken Sound Parkway NW, Suite 300
Boca Raton, FL 33487-2742

CRC Press is an imprint of Taylor & Francis Group, an Informa business

Printed on acid-free paper

International Standard Book Number-13: 978-1-138-55430-6 (Hardback)

Library of Congress Cataloging-in-Publication Data

Names: Lutchman, Chitram, author. | Lutchman, Kevan, author. | Ramakrishna, Akula, author. | Lyons, Charles, author. | Al Hashmi, Waddah Ghanem, author.
Title: Operations excellence management system (OEMS) : getting it right the first time / authored by Chitram Lutchman, Kevan Lutchman, Ramakrishna Akula, Charles Lyons, and Waddah Ghanem Al Hashmi.
Description: Boca Raton, FL : CRC Press/Taylor & Francis Group, 2019.
Identifiers: LCCN 2019020286 | ISBN 9781138554306 (hardback : acid-free paper) | ISBN 9781315148960 (ebook)
Subjects: LCSH: Operations research. | Management science.
Classification: LCC T57.6 .L88 2019 | DDC 658.5/62—dc23
LC record available at https://lccn.loc.gov/2019020286

Visit the Taylor & Francis Web site at
http://www.taylorandfrancis.com

and the CRC Press Web site at
http://www.crcpress.com

CONTENTS

FOREWORD

Your interest in this book might stem from the same challenges I faced several years ago in my quest for operations excellence. I have had various leadership roles in the petroleum refining industry for the past two decades and have been responsible for the safe, reliable, and compliant operations of multiple refineries for the past 10 years. I also have the insight of working for the same company for nearly 30 years. I have had the opportunity to observe the performance of the same plants over long periods and came to the conclusion that with excellent leadership, our plants would perform pretty well. However, when those leaders moved on, the performance of that plant would usually decline. Even if the new leadership were strong, the change in leadership would always precipitate changes in "how we do things," and those changes would be disruptive to our employees and their performance. It was obvious that we needed an operations excellence management system (OEMS) to ensure that we had good, consistent, and sustainable processes to support our excellent employees. An added benefit of a good structured OEMS is the leverage of collaboration – it is far easier to improve how we do things once across multiple facilities than to improve each of those things independently at each facility.

Our OEMS is a structured, organized, and disciplined approach to improving Health, Safety and Environment (HSE), reliability, and business performance. For many organizations, large and small, the goal of operations discipline and operations excellence continues to be elusive.

In this book, the authors set the platform for OEMS as an 8- to 10-year journey to achieve a sustainable OEMS. Having worked with the lead author in the past, it is clear that the book drives home the underlying principle of usable documentation for transitioning the workforce toward an enabled position of doing the right thing the right way every time. We tell people what the right things are in our standards. Disciplined use of procedures and diligent supervision help us ensure consistent delivery of the right things the right way all the time.

When organizations are ill prepared for the OEMS journey, the costly quest for excellence generally fails or subsides to another initiative that is often very difficult to rejuvenate. When organizations embark on the journey of OEMS, they must be well prepared and fully committed to OEMS for the long haul, including the adequate resourcing of people and money. They must also have a thoughtful plan around the way we manage people and organizational changes. The authors emphasize the need for effective change management support to OEMS networks, which are the teams that will lead the implementation and sustainment of operations excellence.

This well-written book easily communicates to users how to achieve success in building and implementing an OEMS, and more importantly, how to get it right the first time. The many principles and practices discussed in this book reflect the collective global experiences of the authors who created and shaped the content of this book based on practical work experiences from organizations that have either implemented or explored implementation of OEMS. The book leverages the learnings from many organizations including my own. The easy-to-read and easy-to-understand concepts and principles discussed in this book takes organizational leaders back to the humble plan–do–check–act (PDCA) cycle as fundamental to successful implementation of OEMS.

... In my view, this book provides a simple pathway for getting OEMS right the first time.

Jim Stump
SVP Operations, HollyFrontier Corporation

This book represents a fourth installment of the lead author – Dr. Chit Lutchman in the series – designed to improve organizational HSE, reliability, and business performance. This book provides an insightful approach to organizations in their quests for operational excellence the first time they seek excellence in business performance. The core principle of the book is the need for a systematic and organized approach toward OEMS with the ultimate goal of doing the right thing the right way every time in the organization. The book begins with the basic premise that we must tell people what the right things are in standards, how to do the right things in procedures and operations discipline in doing so every time through integration, supervision, and appropriate enablement tools and processes.

Across the globe today, many different industry organizations are either seeking to establish OEMS or revisiting unsuccessful attempts at OEMS. Their goals are being able to improve efficiency, business performance, and sustainability. Indeed, this phenomenon is independent of industry type as leaders seek to improve HSE, reliability, and performance and take hold of the reins of sustainable business excellence. There is limited documented work with practical knowledge available today on OEMS to assist organizations in their quest for excellence.

In this book, an attempt has been made by the authors to bridge this gap and provide critical guidance to business leaders on practical approaches to establishing an OEMS. Written for users, the knowledge provided in this book is designed to provide an end-to-end process for organizations of all industries to achieve operational excellence. The best practices and many graphic principles provided in this book are based on practical experiences of the authors from organizations that have pursued excellence in management systems.

Recognizing the challenges of interpretation in documentation, the authors have followed the basic principles of writing for users to convey the key messages and processes to users in a simplified and usable manner.

Built on leading-edge principles, the authors take this OEMS process to the simplest level of following the plan–do–check–act (PDCA) model to achieve excellence. The book emphasizes the importance of learning from others and the need for evolutionary change to the existing management system with the critical need for a shared vision and extreme attention to organizational and people change management. The book provides guidelines on pre- and post-OEMS implementation requirements and strategies to sustain effective OEMS in an organization. Finally, the authors succeeded in translating rich practical experience into simple approaches that really add value to those organizations that are seeking operations discipline and excellence in their business processes. This book provides a structured pathway for organizations to repackage current management systems to achieve excellence or for those at first attempt to get it right the first time.

Carl Brown, PE
Senior Consultant, Kellogg Brown and Root (KBR), Houston, Texas, USA

ACKNOWLEDGEMENTS

This book provides a pathway for organizations to drive operations discipline and achieve operations excellence in business performance. We provide a simple step-by-step process for achieving excellence by leveraging knowledge and expertise derived from years of experience across multiple organizations and industries. Compilation of the knowledge provided in this book would not have been possible without the help of many industry experts.

We acknowledge the many direct and indirect contributors to the content of this book. At the top of the list are Kevin Brown, Robert Boans, Brian Fillmore, John Walker, Marcial Artz, and Steve Gibson. They are a group of experienced and talented operations leaders of HollyFrontier Corporation whose practical knowledge and perseverance have helped shape the content of this book. Included also from HollyFrontier are Tom Shetina and Randy Patton, senior leaders of the organization whose questioning attitudes contributed to the robust content of the book.

Sincere gratitude is also extended to Brian Baron, John More, and Patrick Gareau of Husky Energy's downstream business unit. The indirect contributions of these committed leaders had a positive impact on the overall content of this book. Their knowledge and experiences with operational integrity challenged much of the knowledge and processes presented in this book. We are also extremely grateful to the contributions of Mr. Ahmed Khalil – Director of Fire and Safety at Bahrain Petroleum Company (BAPCO). Involved with OEMS since the early 2000s, Ahmed has been extremely generous in sharing his knowledge and experiences in implementing and sustaining OEMS at BAPCO.

Finally, we sincerely recognize and thank Jim Stump (SVP Operations at HollyFrontier Corporation) for his valued contributions in reviewing the content of this book and providing his feedback. Jim's application of the behaviors associated with high-reliability organizations – questioning attitude, integrity and courage, engagement, increasing knowledge, and a keen sense of chronic unease – has challenged and shaped much of the thoughts and content of this book, for which we are extremely grateful.

To the many unnamed contributors who have impacted us throughout our careers and have shaped our thought and contributions to this book, we are forever grateful.

Dr. Chitram (Chit) Lutchman
Lead author

AUTHORS

Dr. Chitram (Chit) Lutchman, lead author of this book, is a hands-on oil and gas professional with more than 30 years of experience in leading operations management, project management, Health, Safety and Environment (HSE), Process safety Management (PSM), and OEMS. Chit has been the lead author of books published in the areas of project management, EH&S, PSM, and OEMS. He is also the lead author of *7 Fundamentals of Operationally Excellent Management System*. In his work, Dr. Lutchman seeks to bring business intelligence, operations discipline, risk management, reliability, and EH&S under a single umbrella for sustainable excellence in business performance.

Dr. Lutchman is an internationally recognized EH&S and management system professional and has presented at international conferences in the United States, the Middle East, Asia, and the Caribbean. A consummate learner, Dr. Lutchman challenges himself and others to continually improve. He holds DBA, MBA, and B.Sc. (agricultural sciences) degrees. He is also a first-class Power Engineering License holder and is a certified safety professional (CSP) and a Canadian registered safety professional (CRSP).

Dr. Lutchman has provided solutions for improving operations discipline and managing operations risk and HSE solutions to the Emirates National Oil Company (ENOC), Bahrain Petroleum Company (BAPCO), and North American Oil and Gas majors. Today, Dr. Lutchman is regarded among the world's leading experts in EH&S and OEMS.

Dr. Kevan Lutchman has held the part-time role of document writer and editor of technical documents with Safety Erudite Inc. for the past 4 years while completing his degree in medicine. He is a strong advocate of management systems and operations discipline. His interest in workplace health and safety has encouraged him to explore specialization in occupational medicine as a future career, with the intent of proactive intervention in workplace injuries and diseases. Dr. Lutchman has been deeply involved in the first installment of this series having provided editorial, content, and usability reviews of the book titled *7 Fundamentals of an Operationally Excellent Management System.*

At present, Dr. Lutchman continues to focus on developing his career in medicine and looks forward to contributing to further work in the field of occupational health and preventative measures for improving workplace health and safety performance.

Dr. Ramakrishna (Rama) Akula is passionate about business/operations excellence and is committed to helping clients achieve their elusive goals of operations discipline and operations excellence. Dr. Akula is a mechanical engineer with 19 years of experience working for oil and gas organizations in North America, Europe, the Middle East, and Asia.

Dr. Rama gained significant recognition as a subject matter expert in plant management, HSE, organizational strategic performance, operational excellence, and business excellence (quality) managements systems.

Dr. Akula completed his doctorate degree (PhD) in material safety with a focus on improving quality and safety around liquefied petroleum gas cylinders. He is a chartered member in Institution of Occupational Safety and Health (IOSH), UK Chartered Member In Occupational Safety and Health (CMIOSH), holding the title Occupational Safety and Health Practitioner and a Chartered Mechanical Engineer from the Institution of Engineers (India). He also holds an MBA degree in operations, a postgraduate diploma in HSE, and a certificate in disaster management.

An experienced management system assessor, Dr. Rama has been a lead auditor for recipients of prestigious industry excellence awards in the UAE for the implementation of Corporate Excellence programs. Also, Dr. Akula is the recipient of the UAE National Quality Award from American Society for Quality (ASQ) in Middle East and North Africa, and holds a world record on research publications.

Charles (Chuck) Lyons is a business-focused executive with over 30 years of experience delivering results in a broad spectrum of industries across 42 countries. He has expertise in developing, implementing, and managing operational excellence, safety, health, environment, quality, and sustainability at all levels within the organization. Lyons has proven ability to develop and deploy ISO complaint-integrated management systems. He tracks record of driving improvement across the OEMS spectrum. He is an excellent communicator who excels at building strong client relationships to drive business consistency and streamline systems and processes. He possesses diverse QHSE/OEMS experience in both large and small organizations.

Mr. Lyons holds a BSc degree in chemical engineering and an MBA degree from Texas A&M University in 1986 and 2013, respectively, and is a registered professional engineer. He began his career with an international petrochemical company as a process engineer and soon moved into environmental, health, and safety. After 10 years with major petrochemical and refining operators and moving to corporate lead roles, he shifted into the oilfield services business for 4 years working internationally as the corporate lead for HSE and quality for Schlumberger. From there, he shifted back to the major oil and gas operators with Conoco for 4 years before making the jump into engineering, procurement, and construction with KBR as the global executive responsible for quality, health, safety, and environment. Mr. Lyons has spent the last 7 years working in Calgary for oil and gas, power, and utility operators as an executive developing HSE and OEMS systems and managing a diverse operational and asset portfolio.

Dr. Waddah S. Ghanem Al Hashmi has more than 22 years of experience in the engineering, oil, and gas sector. His expertise is in several areas including HSE, management systems, human factors, sustainability, operational excellence, reflective learning for HSE practitioners, safety management as well as environment and energy and resource management.

He is a prolific writer and author of more than 100 technical papers, presentations, and workshops, and has to date authored and coauthored six international published books, one of which was the *7 Fundamentals of Operationally Excellent Management System.*

He sits on several boards including the Energy Institute in the UAE and the UK, and Oil Companies International Marine Forum and Dubai Carbon PJSC in the UAE. He is a Fellow of the Energy Institute (the UK), a member of the Institute of Directors (IoD) (the UK), a member of the American Society of Safety Professionals (ASSP) (the USA), and an Associate Fellow of the IChemE (the UK). He also represents the UAE as Head of Delegation in the TC 283 the International Technical Committee for the ISO 45001 standard.

He is the Hon'ble Chairman of the Energy Institute in the Middle East since late 2017.

1 INTRODUCTION TO OEMS

1.1 The Problem

Overview

OEMS continues to be a goal pursued by many organizations today. Some studies suggest that almost every organization in the energy industry is, in one way or the other, pursuing or exploring the concept of OEMS. The quest for operations excellence and operations discipline is also very vibrant in many other industries. Unfortunately, with few exceptions and regardless of industry, OEMS has been an elusive goal for many organizations.

In this book, the authors provide a structured approach to developing and implementing a sustainable OEMS and a path for getting OEMS right the first time. The knowledge and information provided in this book pulls together collective expertise of ~100 years of experience in the energy industry from the authors and contributors to the content of this book.

The Problem

For many organizations today, a management system (MS) that helps guide the business processes is an absolute necessity. Those that have recognized the value of being disciplined in the application of the MS have surfaced to be top performers within the industry with optimal resource allocation, operating efficiencies, and business performance. On the other hand, a majority of organizations continue to perform at moderate and suboptimal levels for the following reasons:

- Weak understanding and appreciation of the benefits of an organized, structured, and disciplined approach to doing business
- OEMS remains primarily a buzzword where leaders fail to appreciate the real value of operations discipline
- Leaders fail to understand the interrelatedness of all elements of the MS, thereby operating within silos
- The presence of multiple MSs resulting from mergers and acquisition
- Failure to clearly define and communicate the requirements and expectations of the organization
- Placing the responsibility for *doing the right thing the right way every time* on its people without delineating the right things (in standards) and the right way (in procedures)
- Failure to demand operations discipline from its management and workforce

---

OEMS – The Solution?

Many organizations recognize that the shortcomings identified above may be reduced considerably by the application of an OEMS. Indeed, studies suggest that in the energy industry, almost every organization is at some stage in exploring, evaluating, or implementing OEMS.

The challenges, however, for those who have attempted OEMS are as follows:

- Placing the responsibilities for implementing and developing OEMS into the hands of individuals who were
    - Not fully convinced of the need for OEMS, leading to weak buy-in, ownership, and thus stewardship
    - Insufficiently competent for the magnitude of the undertaking
- Inadequate preplanning for the implementation and development of OEMS
- Underestimating the cost associated with undertaking the improvements required for an OEMS
- Unrealistic time expectations for implementation and realizations of the fruits of an OEMS, especially when it comes to an organizational culture change
- Lost focus from a failed first attempt and abandoning the effort
- Failure to make OEMS the central focus of the organization with appropriate recognition

OEMS programs are a long-term commitment organizational transformation project and therefore require a full project management office (PMO) approach, a change management philosophy, and the right number of committed and trained resources at every level of the organization.

1.2 Goals and Objectives of This Book

Goals and Objectives

OEMS and operations discipline have surfaced today in many industries as the basic requirements for stakeholder value maximization and sustainable business performance.

The goals and objectives of this book are to

- Provide organizations a model for successful implementation and sustainability of OEMS.
- Help organizations avoid the frustrations and costs associated with false starts.
- Help organizations apply best practices and successful strategies adopted and applied by early adopters.
- Optimize resources by getting it right the first time while avoiding costly mistakes and rework

Who Should Read This Book?

This book should be read by the following users:

- Business executives and industry professionals
- Health, Safety, and Environmental (HSE) and business professionals involved in implementing and sustaining any type of MSs – HSE, Process Safety Management (PSM), and OEMS
- University students pursuing MBAs and doctoral degrees in business administration

1.3 Overview of Various Types of MSs

Why MSs

MSs provide organizations an organized, structured, and consistent process for managing their business. A MS provides the framework for sustainable development and performance. While there are many MSs available to organizations for stewarding a business – Total Loss Management (TLM), PSM, ISO 9000, and OEMS – the ingredients that lead to sustainable business performance are as follows:

- The right level of detail supporting the high-level requirements and expectations provided in traditional MSs such as ISO 9000, ISO 14001, and ISO 45001 standards, and API 1173
- Consistent interpretation of the requirements and expectations across all entities in the business unit or across all business units of the organization
- Operations discipline in the use and application of standards, procedures, and tools intended to deliver on the requirements and expectations of each element of the MS
    - Boils down to *doing the right thing, the right way, all the time*
    - No shortcuts – we *always* follow the established process/ procedure. Operators must not only take a temperature or pressure reading at a particular frequency but also know *why* they do and *how* corrective actions impact business performance

---

Types of MSs

There is no one silver bullet for business success and performance. Similarly, there is no one MS that can generate exceptional business success. Various types of MSs are found across industries with varying levels of success. Over time, industries have evolved business MSs to leverage those that work best for the industry. Within these industries, corporations have evolved these MSs to differentiate themselves from competitors. The following table provides an overview of the various MSs applied in industry:

MS Type	Overview
Functional MSs	• Functional MSs are typically focused on activities to be managed within each functional team without any major focus on the big picture corporate view • Typically, soloed with little integration of processes and practices • Functional MSs' marketing, finance, and personnel are characterized by force-fitting functional MSs into business plans with little synergy or linkage between them • The lack of a comprehensive, integrated management system (IMS) makes management a much more erratic process than need be • Management of change is essential for the long-term success of the business, and often change initiatives fail due to the lack of integrated and organized systems (Hall, 1994)

Process MSs	• Organizations with process MSs are generally comprised of elements that are mainly core business processes • Process MSs are based on a view that organizations are comprised of a system of interrelated and linked processes • Process MSs involve concerted efforts to map, improve, and adhere to organizational processes (Benner and Tushman, 2003; Chang, 2006)
Centralized governance	• A system of governance that comprises a centralized corporate team within a functional group – typically the HSE functional group • Works best in process MSs
Distributed governance	• A system of governance made up of centers of excellence or business unit/site-specific governance teams • Works well in discrete, siloed business units or facilities within the business unit • Typical of organizations with mature MSs
Operations focused	• MS focused on operating assets only • Functional groups are exempt from meeting the requirements and expectations of the MS
Corporate focused	• The MS includes both functional groups and operating assets of the business • All business activities are subject to the requirements and expectations of the MS • Functional groups are subject to lighter risk-based application of requirements and expectations of the MS

1.4 OEMS – A Definition

Authors' Definition of OEMS

The authors define OEMS as follows:

An OEMS is an *integrated, organized, structured,* and *disciplined* approach to protecting people, the environment, and organizational assets while improving asset integrity, reliability, and business performance.

Variants of OEMS

Organizations differentiate their MSs using varying terminologies to refer to it as follows:

- A series of elements and sub-elements communicating the high-level direction and expectations of the organization
    - The number of elements may vary from 9 to 18 and depending on the industry in which the business is involved
- Detailed requirements and expectations generally defined in one or more standards for each element

The following table highlights variant titles and acronyms used to identify the OEMS of different companies in the oil and gas industry:

Company	OEMS
ExxonMobil	• Operations integrity management system (OIMS) (Exxon Mobil, 2017a)
Chevron	• Operational excellence management system (OEMS) (Chevron, 2019)
Suncor Energy	• Operational excellence management system (Suncor Energy, 2017)
BP	• Operating management system (OMS) (BP, 2017)
Husky Energy	• Husky operations integrity management system (HOIMS) (Husky Energy, 2017)

OEMS Definitions – Oil and Gas Industry

Operations excellence as defined by the oil and gas industry is as follows:

Company	Definition
ExxonMobil	• Exxon Mobil defines operations excellence as follows: • "The Operations Integrity Management System (OIMS) is a cornerstone of our commitment to managing SSH&E risk and achieving excellence in performance" (Exxon Mobil, 2017a)
Chevron	• "Our Operational Excellence Management System (OEMS) is a comprehensive, proven means for systematic management of process safety, personal safety and health, the environment, reliability, and efficiency" (Lohec, 2017)
Suncor Energy	• "To Suncor, operational excellence means operating in a way that is safe, reliable, cost-efficient and environmentally responsible" (Suncor Energy, 2017)
BP	• "Our operating management system (OMS) is a group-wide framework designed to help us manage risks in our operating activities and drive performance improvements" (BP, 2017)
Husky Energy	• Husky Operational Integrity Management System (HOIMS) is designed to generate "Safe and reliable operations" (Husky Energy, 2017)

OEMS Definitions – Other Industries

Operations excellence as defined in other industries as follows:

Company	Definition
Manufacturing	• Manufacturing or operational excellence refers to "efficiency, productivity, and reliability, with minimized downtime and few product failures" • "Manufacturing excellence is not just the best way to do it today, but continually improving to the next level" (Markarian, 2017)
Pharmaceutical	• Manufacturing excellence refers to serving patients by "providing safe and effective medicines without an interruption in supply" (Markarian, 2017)
Construction	• Operations excellence refers to the removal of waste across the entire construction supply chain and become lean with proactive construction companies developing agile network of stakeholders who exchange information and coordinate seamlessly to deliver projects reliably, on time, and under budget (Vaidyanathan and Mundoli, 2014)

1.5 OEMS – A Historical Perspective

Early Introduction of MSs

Early introduction of MSs were designed to improve specific aspects of business performance. MSs have since evolved over time to provide structured and organized ways of doing business. For most organizations, competitive business environments have led to the following:

- Structured approaches to managing the business
- Continuous evaluation and assessment of the MS to improve business performance
- Multiple MSs designed to more effectively manage the business, such as
  - Information MSs
  - Learning MSs
  - Document MSs
  - ISO-based MSs
  - Environmental, health, and safety (EH&S) MSs
  - Process Safety Management
  - Integrity Management System
- Overall improvements in business performance

Benefits of a MS

Some of the benefits derived from MSs are detailed in the following table:

	Benefits
Organizational	• Improved organizational and business performance • Lower employee turnover – greater employee loyalty and retention • Worker productivity improvements • Improved specific assets/aspects of business performance
Manager	• Fewer conflicts and more effective conflict resolutions • Efficiency and consistency in business performance • More effective communication • Clear accountability and performance expectations • More effective budget controls and management
Employee	• Worker expectations clearly defined • Promotes self-assessment opportunities and clearly communicates accountability and performance expectations • Career paths more clearly defined with enhanced job satisfaction

Evolution of MSs to OEMS

Figure 1.1 demonstrates the evolution of MSs toward OEMS.

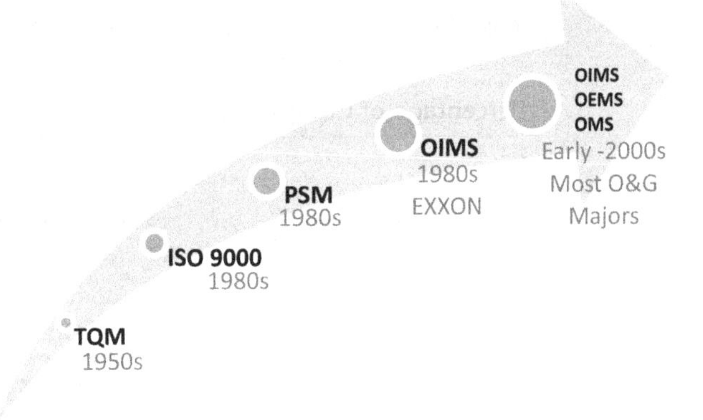

Source: © Safety Erudite Inc. (2019).
Figure 1.1: Evolution of MSs toward OEMS.

- Early introduction of OEMS was led by Exxon in response to the Exxon Valdez incident of 1989.
- The oil and gas industry led the charge toward operations excellence based on the successes derived by ExxonMobil.
- Large-scale implementation of OEMS in the oil and gas industry began in the late 2000s.

The Origin of OEMS

OEMS is a systematic approach to managing the business. Early introduction of OEMS occurred in the oil and gas industry by ExxonMobil in response to the 1989 Exxon Valdez incident in Alaska.

ExxonMobil introduced OIMS in its operations in the early 1990s to prevent future incidents. Since its introduction,

- OIMS has been implemented across its global operations.
- ExxonMobil credits record and world-class environment, health and safety (EH&S) performance to operations discipline and its systematic systems.

(ExxonMobil, 2017b)

1.6 Drivers and Traditional Focus of OEMS in Organizations

Why Organizations Implement OEMS

Ernst and Young (2017), after a review of 30 oil and gas organizations including majors, offered the following reasons why organizations pursue OEMS:

- For HSE improvements, generally after a major incident
- As a proactive response to a major HSE incident in the industry
- For financial and operating performance improvements (including cost reduction)
- To reduce gaps between self and peers
- As a strategic initiative intended to transform the organization

(Ernst and Young, 2017)

Focus on OEMS

Ernst and Young (2017) in the same review identified the following six focus areas of OEMS in the companies reviewed:

Percentage of Companies	Focus of OEMS
80	• Expansive asset reliability
80	• Expansive production efficiency
73	• Management of HSE risks
60	• Operating cost reduction
20	• Focus on culture
13	• Employee retention

Source: Ernst and Young (2017).

Outcomes of OEMS

Ernst and Young (2017) in the same review identified the following six focus areas of OEMS in the companies reviewed:

Percentage of Companies	Outcomes of OEMS
43	Contractor and employee performance improvements in • Recordable injuries • Loss time incidents (LTIs) • Spill reductions

43	• Operating cost reductions
29	• Improvements in asset reliability and availability
29	• Production improvements that may have been linked to asset reliability and availability

Source: Ernst and Young (2017).

1.7 Benefits of an OEMS

Strategic Focus for Value Maximization

For shareholders, an IMS helps mitigate business operations and enterprise business risks and allows the organization to focus on value maximization for its stakeholders.

The following four operational excellence strategic goals provide focus and direction:

- People – Develop the mechanisms to attract, recruit, and retain the best and the brightest in the industry
- Process and personal safety management – Joint focus on process and personal safety management minimizes loss and harm
- Corporate citizenship – Minimize footprints and excel in environmental management performance. Bring value to the communities within which the organization operates, and maintain a strong ethical and moral compass
- Reliability – Develop and implement processes to significantly improve the reliability of the business

Benefits of an OEMS

The primary benefits of an OEMS are summarized in the following table:

	Benefits
Organization	• Stakeholder value maximization reflected in overall organizational and business performance • A motivated workforce that wants to stay • A structured, organized, and disciplined approach to business Consistent and proactive approaches to business • Processes and systems dependence that promotes business continuity • An employer of choice – recruitment of the brightest and the best A shared vision with common purpose, culture, and pride
People	People matter, but processes, systems, and structures exist to allow continuity should they leave Competency – everyone knows what is required of them to maintain safe and reliable operations Mature networks to sustain and continually improve business performance People are empowered with clearly defined roles and set boundaries – decision rights are defined and never second-guessed

EH&S	• Thoughtful risk management across the company • Learning from internal and external incidents • Processes are consistent where they need to be and customized where they need to be, no matter what they are sustained and improved • Silos are eliminated – sharing and collaboration is entrenched
Reliability and integrity	• Assets are properly cared for with fewer unplanned outages or loss of containment

EH&S and Financial Performance

EH&S and financial performance of companies that have implemented OEMS or its variants are compared in the following table. The period of data comparison is 2003–2012 where data was available and accessible.

Company	OEMS – Years Implemented	Average ROCE (%)	Average Net Income ($ Billion)	Average CRIF	Average ERIF
ExxonMobil	27	26.17	34.36	0.06	0.06
Chevron	17	20.4	17.69	0.38	0.36
Suncor Energy	5	2.34	13.46	0.90[a]	0.62[a]
BP	10	14.48	16.91	0.57	0.37
Husky Energy	10	16.92	2.07	1.23[b]	0.73[b]

[a]2007–2012.
[b]2009–2012.

1.8 Relationships between OEMS, PSM, and Occupational Health

Cumulative Relationships

The relationships between OEMS, PSM, and occupational health (OH) are detailed as follows:

- Relationships are cumulative.
- OH is integrated into PSM to produce the organization's safety management system (SMS).
- PSM is baked into OEMS.
- Excellence in business performance is the outcome of OEMS, operations discipline, the right structure, and an effective SMS.

Figure 1.2 demonstrates the relationships between OEMS, PSM, and OH.

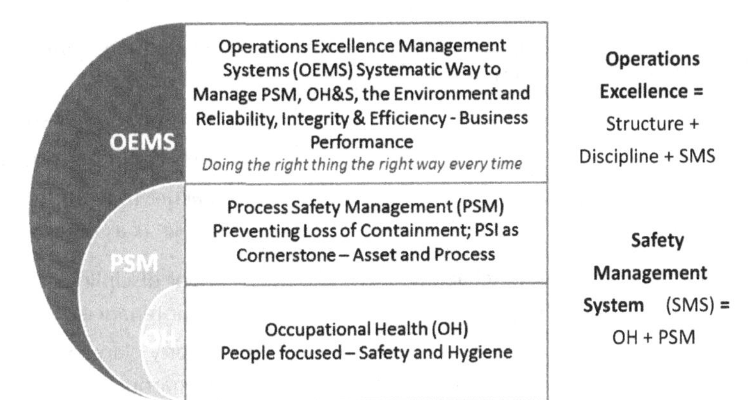

Source: © Safety Erudite Inc. (2019).
Figure 1.2: Cumulative relationships between OEMS, PSM, and OH.

1.9 Toward a Successful and Sustainable MS

Integrated Process MS

An integrated process MS that appears to have excelled and demonstrated sustainability over time is what industry today terms an OEMS and its variants (discussed earlier). According to Ernst and Young (2017), continued and sustainable results in oeprations excellence (OE) across industries were driven by the following integrated business attributes:

- Focused attention to environment, health, safety, and quality
- Integrated approaches to business planning
- Effective operating models
- Successful contractor and supplier management
- Improvements in asset reliability and integrity
- Improvements in cost-effectiveness and efficiencies
- The authors contend that consolidation of multiple inherited MSs toward a single, standardized, and consistent MS throughout the corporation is also essential for success

Organizational Failures in OE

Ernst and Young (2017) advised further that failures in OE were constrained by the following:

Business Attributes	Details
Integrated business planning	• Inadequate – or lack of – leadership, sponsorship, and maturity • Ineffective strategy and framework to support the business planning process • Cumbersome and difficult-to-use processes for translating strategy into operations
Operating models	• Siloed design architecture • Unclear decision governance • Insufficient and inefficient performance management • Weak operations discipline – standardization, policy enforcement, and work practices

Suppliers and contractors	• Inadequate contractor prequalification and execution of contracts • Contractor and supplier risks not properly understood • Weak field performance management of contractor • Failure to learn from prior projects – regards all projects as different
Asset reliability and integrity	• Poor discipline in predictive and preventive maintenance • Poorly planned and executed shutdown and turnaround • Operations readiness planning ineffectiveness • Playing catch-up – over-abundant maintenance activities • Inadequate date and information management
Cost efficiency	• Stakeholders unaware of or do not understand cost drivers across value chain • Production management optimization challenges • Simulation and business modeling inefficiencies
Multiple MSs – lack of standardization and consistency	Inadequate standardization and lack of consistency result in the following: • Multiple fragmented cultures and sub-par business performance • High and low cost of production entities • Internal competition versus collaboration

Organizational Success in OE

Ernst and Young (2017) advised further that successes in OE were enhanced by the following:

Business Attributes	Details
Environment, health, safety, and quality	• Operational risks are understood and ranked • Full involvement – everyone is engaged and involved in a journey to eliminate incidents (including environmental incidents) across the corporation • Strong quality assurance and control measures are in place to deliver a zero-defect approach to product quality
Integrated business planning	• Business strategies are aligned with the corporate vision • Tactical and annual plans are aligned with the strategic plan and corporate vision • Plans are properly resourced, supported, and executed
Operating models	• An integrated approach to people, processes, and systems is maintained to support the business • Operating models are designed to achieve optimum efficiency

Suppliers and contractors	• Contractors and suppliers are valued stakeholders who are integrated into the business process • Contractors and suppliers contribute positively to business performance • Performance is actively managed, consistent with deliverables and expectations
Asset reliability and integrity	• Core principles of asset care management are embraced and institutionalized so as to • Identify and eliminate potential asset failures • Learn from failures and establish sustainable corrective actions • Focus on asset life cycle management – design through decommissioning
Cost efficiency	• Stakeholders are vigilant in their pursuit for cost improvements and continuous improvements in all aspects of the business • Focus on getting it right the first time to eliminate rework • Enablement and optimization tools and technologies are available to increased production and outputs
A single MS – standardized and consistent work practices	• Adoption of the lowest cost production process • Internal collaboration and sharing of knowledge • Continuous improvements

OE Success and Cultural Attributes

The authors concur with Ernst and Young's findings and recommendations regarding success in operations excellence. An integrated process MS is by far more effective than a functional MS in today's diverse and complex business environment.

In their study, Ernst and Young (2017) advised further that the cultural attributes for success in OE included the following:

- Strong leadership
- Engaged personnel
- A clear focus
- Leveraged technologies
- Operations discipline (added to the list by the authors)

Details of each attribute are provided as follows:

Cultural Attributes	Details
Strong leadership	• Transformational leadership behaviors are demonstrated throughout the organization – inspiring hearts and minds • A shared vision – realistic and achievable • Developing the entire organization to higher levels of performance • Performance driven

Engaged personnel	• All workers are involved and engaged in the work process and deliverables • Workers are continually seeking and driving improvements • Clear recognition of alignment of work, and how individual performance impacts the performance of the entire organization • Accountability – benevolently holding each other accountable throughout the organization
Clear focus	• Operational excellence and sustainable performance of the organization shared by all internal stakeholders • Empowered workforce • Timely responses to the business environment
Enabled organization – leveraged technologies	• The organization is enabled by right-sized and fit-for-purpose technologies • Communication • Performance management • Data transformation into information and knowledge • Information management promotes timely decision-making and effective sharing of knowledge
Operations discipline	• All workers are vigilant and disciplined in the application of policies, standards, procedures, and work practices • Shortcuts and unsafe work are not tolerated

1.10 Fundamentals of an OEMS

Seven Fundamentals of an OEMS

An OEMS is not sustainable unless the seven fundamentals are recognized and applied in a consistent manner across all business units and functional groups of the organization. Lutchman, Evans, Ghanem, and Maharaj (2015) listed the following fundamentals of an OEMS:

1. Leadership commitment, motivation, and accountability
2. Establishing the required elements of the MS
3. Establishing the baseline to understand where the organization sits in relation to its desired state
4. Following a plan–do–check–act process for managing work
5. Auditing for compliance and conformance to the requirements of each element standard
6. Closing the gaps and maintaining operations discipline in doing so
7. Continuous improvements and shared learning

Lutchman et al. (2015) advised that success in implementing and sustaining OEMS requires a systematic approach to addressing each of the fundamentals listed above.

Operations Discipline

The authors regard *operations discipline* as a critical component of an OEMS. Operations discipline refers to executing work consistent with the following:

- Requirements and expectations of established standards or regulations where applicable
- Established procedures, processes, and work practices
- *Doing the right thing the right way every time*

1.11 Steps in Implementing an OEMS

Steps in Implementing an OEMS

The steps involved in implementing an OEMS are identified in Figure 1.3. Careful attention to each step is required to ensure a sustainable process. The key steps in the process are as follows:

1. Establish and share vision – Setting the organization on the path by sharing the OEMS vision
2. OEMS elements – Establishing the corporate requirements and expectations in policies and standards and to achieve the vision
3. Establish baseline – Internal benchmarking relative to requirements and expectations
4. Implement gap closure strategy – Built within the annual and strategic plans
5. Sustain – Monitoring, reviewing, and correcting deficiencies
6. Continuous improvement – The organization must be adept at learning, innovating, and adjusting to the business environment

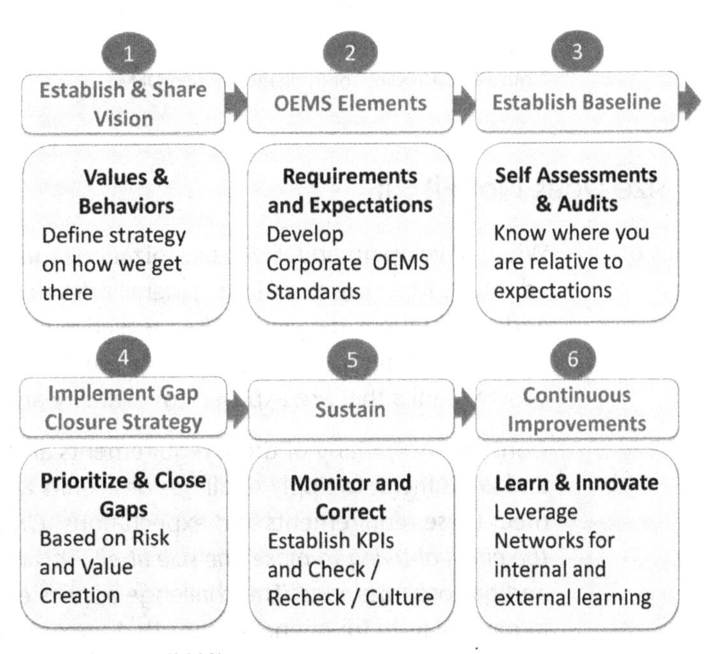

Source: © Safety Erudite Inc. (2019).
Figure 1.3: Steps in implementing an OEMS.

Timeline for Implementing an OEMS

Implementing an OEMS is not a short-term venture. Significant time and effort are required over approximately 5–8 years for effective implementation and sustainment of an OEMS. The timeline for implementing an OEMS is summarized in Figure 1.4. The duration can be increased or decreased based on the following:

- The organization's ability to avoid a *false start*
- The amount of up-front work completed by the corporate organization
- The effectiveness of the organization or team charged with responsibilities for implementation and imbedding the new culture

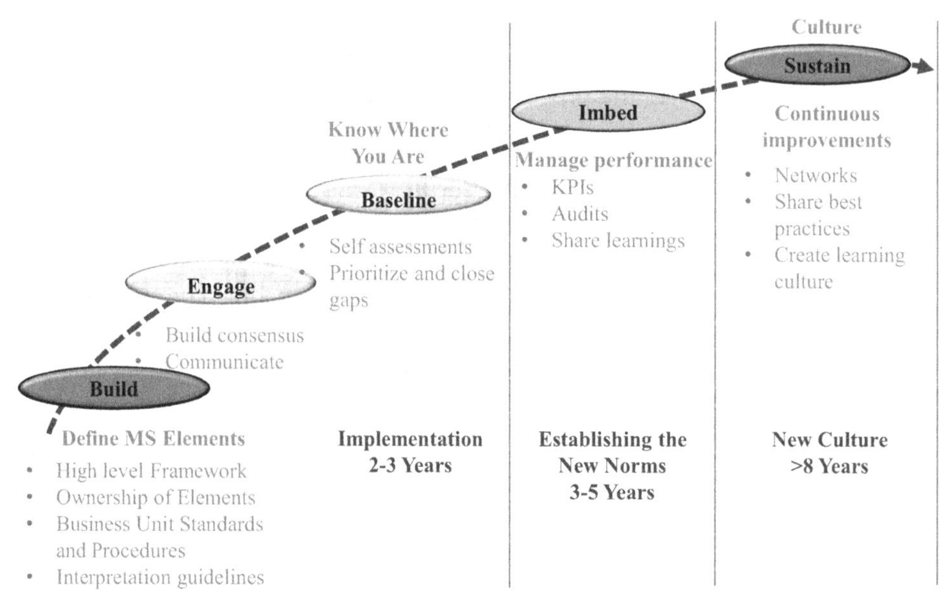

Source: © Safety Erudite Inc. (2019).
Figure 1.4: Timeline for implementing an OEMS.

1.12 One Size Does Not Fit All

Avoiding Pitfalls of Early Adopters

When developing an OEMS, organizations must seek to avoid the pitfalls of early adopters. Most organizations generally develop a set of corporate requirements and expectations detailed in the respective element standard. In most instances, these requirements and expectations are designed for operating assets and business units that are exposed to reliability and EH&S challenges and risks.

Consequently, many of these requirements and expectations are not applicable and are difficult to apply to all business units and functional groups required to meet these requirements and expectations. *Organizations must guard against the pitfall of trying to make one size fit all.* This is particularly so in integrated oil and gas organizations. The challenge applies equally to other industries and, in particular, to functional groups that are common to most organizations and across industries.

Figure 1.5 provides an overview of the concept that one MS does not fit all.

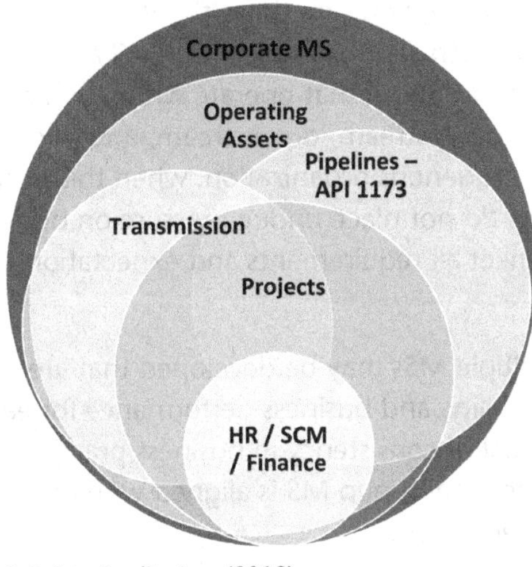

Source: © Safety Erudite Inc. (2019).
Figure 1.5: One size does not fit all.

OEMS and Homogenous Organizations

Homogenous organizations refer to those that focus on a single-core business within a particular market segment. For example, in the oil and gas industry, homogenous organizations are those that operate within defined sectors of the industry such as pure upstream, downstream, or midstream.

- The corporate MS is built around core operations – aimed at reducing operating and enterprise risk and improving asset reliability.
- A goal of standardization in business practices and processes is sought across all assets.
- Functional groups may develop an MS that is aligned with the corporate MS requirements and expectations.

OEMS and Nonhomogenous Organization

A nonhomogenous organization refers to one involved in several core businesses that

- Are not within the same industry or industry sector, for example, oil and gas production, hydrocarbon transportation, and power generation.
- May focus on multiple core businesses within a particular industry sector. For example, in the oil and gas industry, an organization operating in the midstream sector of the industry may be involved in power generation and transmission, gas and liquid collection and transportation, and utility supply to cities and organizations

When this occurs, it is imperative that corporation does not place undue pressure on business units and functional groups to meet all requirements and expectations of each element of the MS similarly. The focus should be the following:

- A corporate MS is developed that is supported by multiple business unit-specific MSs that are designed to manage EH&S, reliability, and business performance for each business unit or core business area.
- A goal of consistency in business practices across business units or core business areas is sought.
- Functional group MS is aligned with the corporate MS requirements and expectations.

OEMS and Integrated Organizations

Integrated organizations are those that focus on multiple core businesses within the industry. For example, in the oil and gas industry, integrated organizations are those that operate within two or more sectors of the industry (i.e., within the upstream, downstream, and midstream sectors). As is the case with nonhomogenous organization, when this occurs, it is imperative that corporations do not place undue pressure on business units and functional groups to meet all requirements and expectations of each element of the MS similarly.

- Multiple MSs may be developed that are designed to manage EH&S, reliability, and business performance for each business unit.
- A goal of consistency in business practices across business units is sought.
- Functional group MS is aligned with the corporate MS requirements and expectations.

OEMS – Homogenous versus Nonhomogenous Organizations

The following table provides a simple comparison between the approaches to MS development for homogenous and nonhomogenous organizations:

Criteria	Homogenous	Nonhomogenous
MSs	Integrated single corporate MS	Multiple integrated MS aligned with the corporate MS
Goal	Standardization of work practices	Consistency in work practices among business units; standardization within business units
Documentation	Corporate policies, standards, and procedures for use by all assets	Corporate policies and standards – flexibility in procedure use within business units
Functional organizations (e.g., HR/SCM/finance, etc.)	Aligned with corporate MS	Aligned with corporate MS

1.13 How We Achieve Operations Excellence

Achieving Operations Excellence

A lot of preplanning work is required to achieve operations excellence. Preplanning efforts are required in the following areas:

- Identifying the appropriate and correct elements of the organization's MS and appropriate requirements and expectations
- Developing and applying the right model for operations discipline

1.14 Is OEMS Applicable to All Industries?

Industry Application of OEMS

Consistent with the definitions provided by the authors as well as those provided by specific organizations, OEMS is about being structured, organized, and disciplined in the execution of work.

- These tenets are applicable across any industry.
- Most industries have been applying the principles of operations excellence and operations discipline in pursuit of improving business performance.

References

BP (2017). BP's Operating Management System. Retrieved June 09, 2017 from www.bp.com/en/global/corporate/sustainability/operating-responsibly/managing-risk.html.

Benner, M. J., and Tushman, M. L. (2003). Exploitation, exploration, and process management: The productivity dilemma revisited. *Academy of Management Review*, 28, 238–256.

Chang, J. F. (2006). *Business Process Management Systems Strategy and Implementation.* Taylor & Francis.

Chevron (2019). Operational Excellence. Retrieved July 24, 2019 from: https://www.chevron.com/about/operational-excellence.

Ernst and Young (2017). Driving Operational Performance in Oil and Gas. Retrieved August 03, 2017 from www.ey.com/gl/en/industries/oil---gas/ey-driving-operational-performance-in-oil-and-gas-1-low-oil-price-highlights-the-need-for-operational-excellence.

ExxonMobil (2017a). Operations Integrity Management System. *Chairman's Message.* Retrieved June 09, 2017 from http://cdn.exxonmobil.com/~/media/global/files/energy-and-environment/oims_framework_brochure.pdf.

ExxonMobil (2017b). Operations Integrity Management System. *Chairman's Message.* Retrieved June 09, 2017 from http://corporate.exxonmobil.com/en/community/corporate-citizenship-report/about-this-report/exxonmobil-operations-integrity-management-system.

Hall, D. (1994). Introducing Second Generation Management. Journal of Small Business and Enterprise Development, 01/1994, Volume 1, Issue 1. Retrieved August 04, 2017 from EBSCOHost Database.

Husky Energy (2017). Community Report 2015. Retrieved June 09, 2017 from http://huskyenergy.com/downloads/abouthusky/publications/sustainabledevelopment/CommunityReport2015/Husky-Community-Report-2015.pdf.

Lohec, W. (2017). Operations Excellence Management System. *Effective Risk Management Help us Operate Reliably, Efficiently and Most Importantly Safely.* Retrieved June 09, 2017 from www.chevron.com/-/media/chevron/PDF-Reports/About/operational-excellence-oems.pdf.

Lutchman, C., Evans, D., Ghanem, W., and Maharaj, R. (2015). *7 Fundamentals of an Operationally Excellent Management System.* CRC Press.

Markarian, J. (2017). Innovating for Manufacturing Excellence. *Pharmaceutical Technology Europe; Cleveland.* Retrieved June 09, 2017 from ProQuest Database.

Safety Erudite Inc. (2019). The Operations Excellence Management System Provider. Retrieved June 02, 2019 from www.safetyerudite.com

Suncor Energy (2017). Operations Excellence. Retrieved June 09, 2017 from www.suncor.com/about-us/operational-excellence.

Vaidyanathan, K., and Mundoli, R.S. (2014). Achieving Operational Excellence in Construction Projects through Process and Technology Alignment. Retrieved June 09, 2017 from www.academia.edu/people/search?utf8=%E2%9C%93&q=%22operations+excellence+in+construction%22.

2 MODELS FOR IMPLEMENTING OEMS

2.1 Implementation Models

Overview

In today's business environment, organizations of all sizes generally apply a management system (MS). While this is hugely beneficial, the effectiveness of the MS often differentiates good and great business performance. As competition increases and the business grows, the quest for performance improvements increases, as does the search for operations excellence and operations discipline.

Organizations view OEMS as an astronomical undertaking and often avoid change until the last minute. Leaders who understand the value of an MS will often set in motion steps to develop and OEMS at the earliest opportunity, thereby allowing them to establish the desired culture sought for the business. This is done with an understanding that cultural change is extremely difficult and very time-consuming – *the way we do business.*

Transitioning an organization's MS to an OEMS requires consideration for the following two implementation models:

1. Model 1: Fresh start – a blank slate
2. Model 2: Upgrading the existing MS

For organizations unfamiliar with OEMS, the preferred model tends to be Model 1. Large homogenous and nonhomogenous organizations may often seek to apply Model 2. In this chapter, the authors explore each of these models and the respective major activities involved.

2.2 Model 1: Fresh Start – A Blank Slate

Fresh Start – Overview

Model 1 – Fresh start refers to exactly the creation of an entirely new MS. It requires replacing the existing MS with a new MS. This does not mean every aspect of the MS is replaced. Rather, once the requirements and expectations of the new MS are developed, components of the existing MS which meet expectations are retained and sustained.

Implementation Model

Major activities associated with Model 1 are provided in Figure 2.1.

- The chain of events or activities may vary within the model up to the point of implementation – i.e., concurrent activities may occur.
- Post-implementation, the chain of events or major activities are sequential.
- Throughout the process, leadership commitment, involvement, and stakeholder feedback are essential for getting it right the first time.
- The duration of the process may vary based on the approach to implementation.
- A transformational approach to leadership is likely to get everyone on board much quicker than an autocratic or democratic approach – inspires hearts and minds.

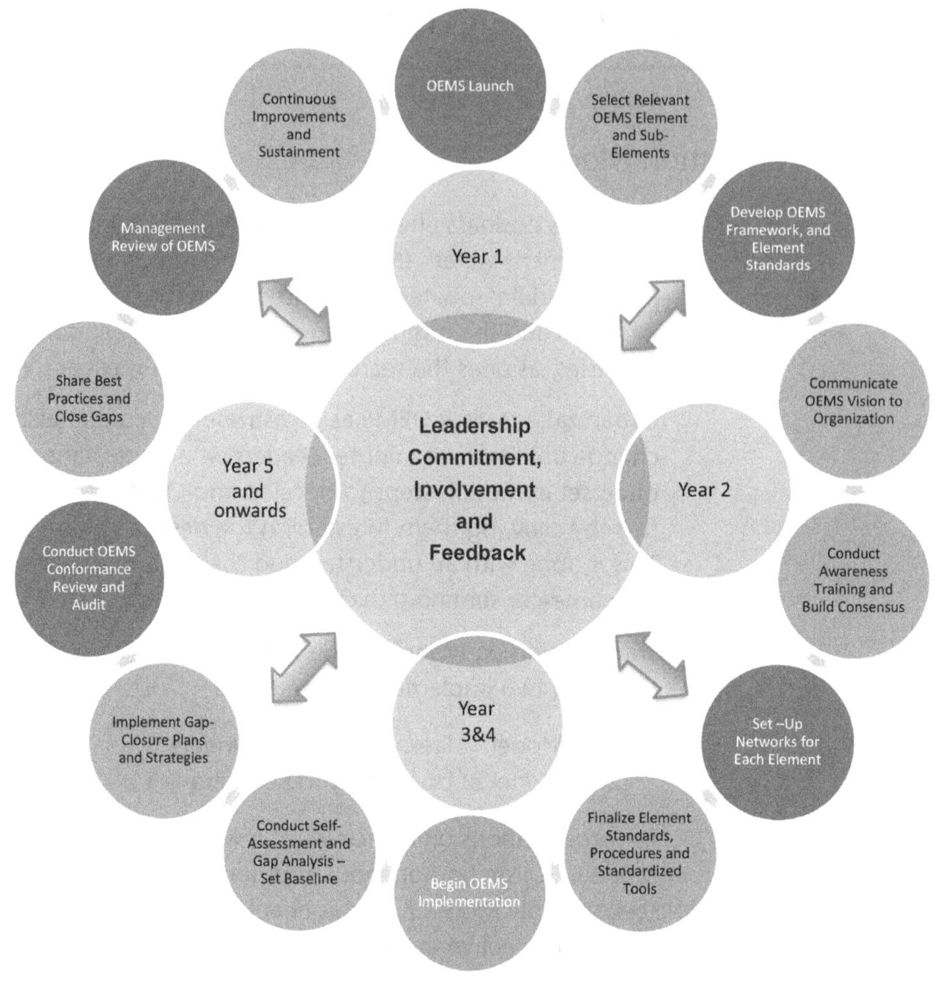

Source: © Safety Erudite Inc. (2019).

Figure 2.1: Model for achieving operations excellence.

2.3 OEMS Launch

OEMS Launch/Kickoff

Launching OEMS is a senior leadership responsibility that requires the following:

- A shared vision and full alignment of all leaders across the organization
- Visibility of leadership in support of the OEMS launch and long-term commitment
- Verification that the organization is ready for OEMS
- Readiness to
    - Transition the organization from people dependence to processes and systems dependence
    - Commit adequate resources to implementing OEMS while recognizing returns on investment will not be received until sometime in the future
    - Leaders are prepared to work toward the greater good of the organization versus vested self-interest

Selecting OEMS Elements and Sub-elements

Selecting an OEMS elements and sub-elements is a leadership exercise that requires a review and assessment of industry peer's MSs. The approach involves the following:

1. Identify industry peers involved in similar business to your organization.
2. Create a spreadsheet allowing comparison of selected peers to identify common elements and sub-elements.
3. Select elements and sub-elements applicable to your business and operations.
4. Right-size elements and sub-elements, repackaging as required to ensure all aspects of your business and operations are covered.

Figure 2.2 provides a simple schematic of the process.

Oil and Gas Industry	Common Elements and Sub-Elements	Your Oil and Gas Corporation
Exxon Mobil Chevron Suncor Energy BP Husky Energy Valero Bahrain Petroleum Company (BAPCO)	1. Leadership Commitment 2. Safe Operation 3. Reliability and Integrity Management 4. Hazard and risk management 5. Management of Change 6. Environmental Management and Regulatory Compliance 7. Incident Management 8. Training and Competency Assurance 9. Document and Information Management 10. Project 11. Emergency Management 12. Contractor and Supplier Management 13. Performance Review	Elements and Sub-Elements, rightsized to meet the business needs of your organization

Source: © Safety Erudite Inc. (2019).
Figure 2.2: Process for identifying elements and sub-elements.

2.4 Develop OEMS Framework and Element Standards

What Is the OEMS Framework?

The OEMS framework is a simple high-level document (brochure) that provides a bird-eye view of OEMS, and its requirements and expectations. The framework is often a communication document to internal and external stakeholders, and it provides an overview of the following:

- The corporate vision for the organization and how OEMS helps us get there
- The OEMS vision
- Elements and sub-elements of the MS
- Corporate owners of the respective elements
- The implementation plan with major milestones and activities as defined in Figure 2.1
- A question-and-answer section

How to Develop Framework

Developing the framework is an exercise best handled by the communications department, or with their input. The following considerations should apply when building the framework:

- This process may require several revisions before getting it right
- Seek professional input to ensure quality work and output
- Leverage pictures showing teamwork and a united effort
- Refer to existing processes where the organization has already been meeting the requirements and expectations of the MS
- Get it right – be realistic in requirements, expectations, and the implementation plans

Where to Locate the Framework

Locate the framework where it can be easily accessed by all stakeholders. Consider the following for locating the framework:

- All levels of the organization, including frontline workers, should have easy access to the framework.
- Locate prominently on the intranet where it is easily accessible by all internal stakeholders.
- Provide external stakeholders access by locating it on the corporate website.
- For stakeholders without computer access, provide hard copies easily accessible in meeting rooms, lunch rooms, etc.

What Are the OEMS Element Standards?

OEMS element standards are required for each of the respective elements. Each element standard provides the organizational requirements and expectations for that element and defines organizational management of the respective element. These standards should

- Approved by the highest level of authority in the organization – the President/CEO.
- Simple, accurate, and easy to interpret.
- Finite with a life of at least 3 years, which gives the organization time to meet the requirements and expectations before changes are made.
- Provide only the "Whats."
    - "Hows" are not included in these standards.

How to Develop the OEMS Standards

When developing the OEMS standards, it is imperative that leadership involvement and buy-in is available to ensure ownership. Developing these standards is an intensive, time-consuming process for which most senior leaders do not have the time to devote. In view of this, when developing the OEMS element standards, the following should apply:

- Consult with the element owner to determine minimum expectations and requirements of each standard.
- Seek external support and expertise to develop the initial versions of the OEMS.
- Review each standard with internal subject matter experts (SMEs) to ensure the following:

- Getting buy-in and support of for the requirements and expectations defined in each standard
- Realism in requirements and expectations
- Ownership among key stakeholders who may be responsible for implementation of these standards and meeting the requirements and expectations defined in them
- Test run the requirements and expectations defined in the standard across a key stakeholder group
- Occasionally, the OEMS network may be the key stakeholder group for test running the standard before finalizing them

---

Where to Locate the OEMS Standards

Locate the OEMS standards where they may be easily accessed by all stakeholders. Consider the following for locating the OEMS standards:

- All levels of the organization, including frontline workers, should have easy access to applicable OEMS standards.
- Locate prominently on the intranet where it is easily accessible by all internal stakeholders.
- For stakeholders without computer access, a binder with hard copies that is easily accessible when required should be provided.

Assigning Ownership of OEMS Elements

Ownership for each OEMS element is assigned to senior leaders of the organization. Occasionally, ownership of several OEMS elements may reside with a single leader. Ownership of OEMS elements is determined based on the following:

- Functional or departmental role – for example, human resources may assume ownership of personnel training and competency assurance management.
- Hierarchical power of the leader – for example, the Senior Vice President of Operations may assume ownership of elements associated with maintaining continuous and reliable operations.
- SME – for example, safe operations, emergency management and other safety-related elements may be owned by the leader of the environment, health, and safety (EH&S) organization.

Communicate the OEMS Vision to Organization

Communicating the OEMS vision to organization is the responsibility of senior leadership of the organization. The CEO and their executive team should perform such communication. Such communication should consider the following:

- Share vision with all levels of the organization.
- Identify attributes of the organization 5 years ago; where we are today and where OEMS can take us.
- Focus on benefits to be derived by all stakeholders and overall health and safety and business performance.
- Recognize that OEMS is a journey that shall generate some instantaneous benefits in some areas and increase workloads in others.
- Provide an idea of the time involved and the commitments of the leadership teams of the organization.

- Be prepared to answer questions regarding why this journey is important – failing to implement OEMS exposes the organization to the risk of falling behind peers.

Conduct Awareness Training and Build Consensus

Awareness training is generally communicated to the entire organization in much greater details than sharing of the vision. Such training should address the following:

- What will be changing and how soon will these changes occur?
- Explain why these changes are required.
- Answer the *WIFM* (what's in it for me) question for stakeholder groups and individuals where applicable.

By doing this awareness training, the organization builds required support and consensus for moving forward and implementing required changes.

2.5 Setup and Activate Networks for Each Element

What Are OEMS Element Networks?

OEMS element networks are groups of SMEs associated with each element assembled by leadership of the organization. Networks are designed to accept ownership for the respective element and be responsible for implementing the requirements and expectations detailed in the element standard. Lutchman, Evans, Maharaj and Sharma (2013) advised networks will typically comprise the following:

- A network leader – typically a corporate organization individual
- A corporate representative where available
- Business unit representative or SME

Figure 2.3 provides an overview of a typical OEMS network structure.

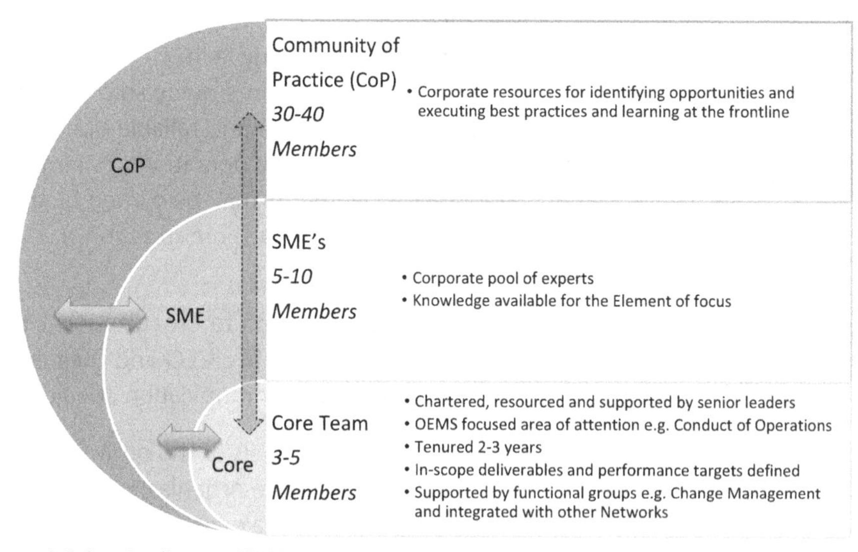

Source: © Safety Erudite Inc. (2019).
Figure 2.3: Typical OEMS network structure.

Refer to *Process Safety Management: Leveraging Networks and Communities of Practice for Continuous Improvements* (publisher – Taylor & Francis) for more details on networks.

Setting Up Networks

When setting up OEMS element networks, careful attention to the skills and behaviors of each network member and the role he/she plays is required in the selection process. Selection criteria for each role are detailed in the following table:

Representative	Selection Criteria
Network leader	Network leaders shall demonstrate the following: • Strong non-authoritative leadership skills • Decision-making capabilities • Strong abilities to remove obstacles and interface with senior leaders of the organization • Comfort with technology; able to lead virtual teams • Ability to create strong teams • Ability to lead and manage change – understands the principles of change management and the 7 Levers of Change • A respected reputation within the organization
Corporate representative	Corporate representatives are required to demonstrate the following capabilities: • Ability to clearly articulate opportunities for standardization where applicable and drive consistency and standardization for the corporation
Business Unit (BU) member	Business unit members shall demonstrate the following: • Strong SME capabilities • Unbiased representation, seeking to deliver collective corporation value versus business unit value • Ability to work independently and take on new challenges and responsibilities with little supervision • Excellent people skills – ambassador for the element (training/coaching, etc.) • Excellent team skills • Strong presentation skills • Comfort with technology and virtual teams • A *doer* – willing to give more than required (consider generation Y personnel)

Finalizing Element Standards, Procedures, and Tools

During the process, the OEMS element network is responsible for finalizing the element standard (when tasked to do so) and developing procedures/processes and tools for meeting the requirements and expectations of the standard across the organization and respective business units.

This process may involve the following activities:

- Collecting and collating all processes and tools used across the entire organization for meeting the requirements and expectations of the standard
- Where multiple business units are available, one can be reasonably assured that there will be multiple processes and tools for achieving the same requirements and expectations

- What this means is that there will be a high-cost and a low-cost process for meeting the same deliverables
- Development of supporting standards, procedures, and tools acceptable to all business units
- Seeking consistency versus standardization among homogenous business units
- Seeking external knowledge to improve internal processes when such knowledge does not reside within the organization
- May be responsible for implementing procedures and tools developed to meet the OEMS element requirements and expectations
- May provide SME support and guidance when not responsible for implantation activities

2.6 Begin OEMS Implementation

Implementing OEMS

Implementing OEMS may take several approaches as detailed below:

- Approach 1 – Full implementation of all elements simultaneously
- Approach 2 – Implementation of elements that are considered priority by the organization
- Approach 3 – Piloting of all elements of OEMS across one business unit, then applying learnings across all other business units during implementation

Regardless of the approach adopted, there are advantages and disadvantages as described in the following table:

Approach 1 – Full Implementation of all Elements Simultaneously	
Advantages	• Time to full implementation is reduced. • Organization learns and develops as a single entity. • Interrelationships among elements are seamless.
Disadvantages	• OEMS is a costly undertaking for the organization. • The resources loading and demand on the organization can be high. • The organization experiences a steep learning curve.
Approach 2 – Implementation of Elements That Are Considered Priority by the Organization	
Advantages	• The resource burden on the organization is reduced to more manageable levels. • Learnings can be shared during the implementation of non-priority elements. • The organization addresses known gaps and areas of weaknesses in the business.
Disadvantages	• The duration of implementation can be long. • All elements of OEMS are interrelated; implementation may not be seamless with non-priority elements.

Approach 3 – Piloting of all Elements of OEMS across One Business Unit, Then Applying Learnings across All Other Business Units during Implementation	
Advantages	• The resource burden on the organization is reduced to more manageable levels. • Key learnings can be shared during the implementation of OEMS across other business units. • Success breeds success; motivation may be high when implementing OEMS across other business units.
Disadvantages	• The duration of implementation can be long. • Business units may not all be the same, so learnings may not work similarly.

2.7 Conduct Self-Assessment and Gap Analysis – Set Baseline

Conducting a Self-Assessment and Gap Analysis

The scope and magnitude of work associated with implementation of an OEMS is defined by the gap analysis. The gap analysis comprises of the business undertaking a comparative assessment of current processes relative to requirements and expectations defined in each element standard.

The self-assessment and gap analysis

- Are performed by element owners' representatives (generally the element network representative) assigned for each operating or business unit of the organization.
- Assess requirements and expectations status on a continuum from regressive to excellence for each of the following criteria:
  - Leadership commitment
  - Risk management
  - Performance outcomes
  - Implementation effectiveness
- Allocate a maturity score for the requirement based on the criteria identified above.
- Develop the baseline assessment for each element collectively providing the baseline assessment for the entire MS.

2.8 Implement Gap Closure Plans and Strategies

Categorizing Gaps for Closure

Gaps may be categorized as per the following table:

Gap Categorization	What We Mean
At-risk gaps	• These are gaps in the MS that exposes the organization is unacceptable and high risk • May be the result of poorly implemented high-criticality work

Low-hanging fruits	• Gaps that may result from inadequate implementation of average criticality work • May not require a huge amount of effort to close the gaps identified
Best practices	• While not technically a gap, best practices are opportunities for sharing business practices and knowledge across the entire organization as applicable

Figure 2.4 provides a typical approach for categorizing gaps identified.

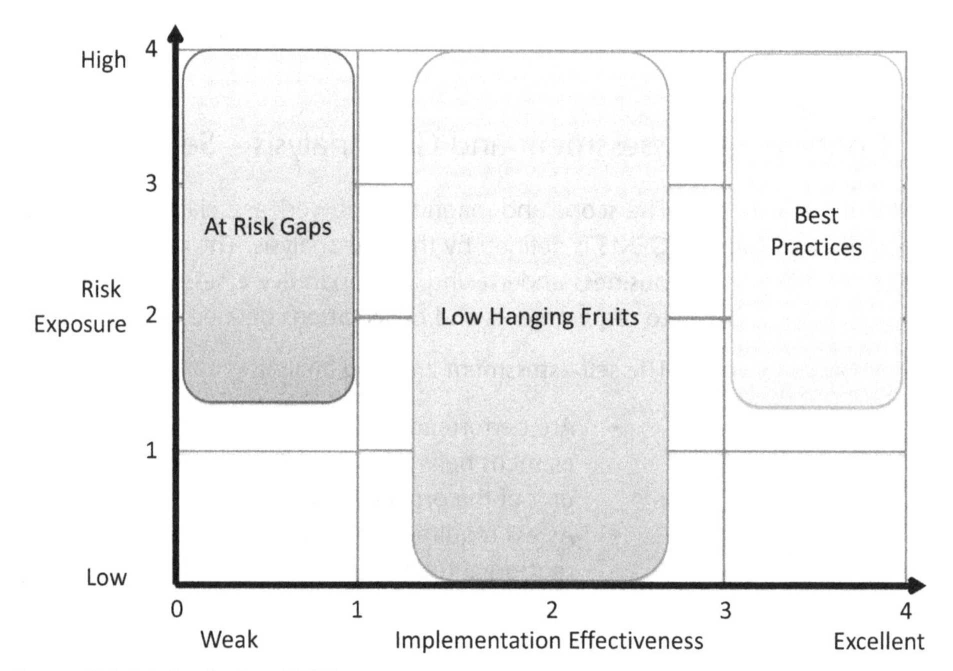

Source: © Safety Erudite Inc. (2019).

Figure 2.4: Typical approach for categorizing gaps identified.

Prioritizing Gap Closure Plans and Strategies

Addressing all gaps together is resource prohibitive. A process is required for prioritizing gaps to be closed in a sustainable way. Gaps may be prioritized based on the following criteria:

- Risk reduction
- Value creation
- Complexity and effort required to implement solutions

Figure 2.5 provides a model for prioritizing gap closure plans and strategies developed.

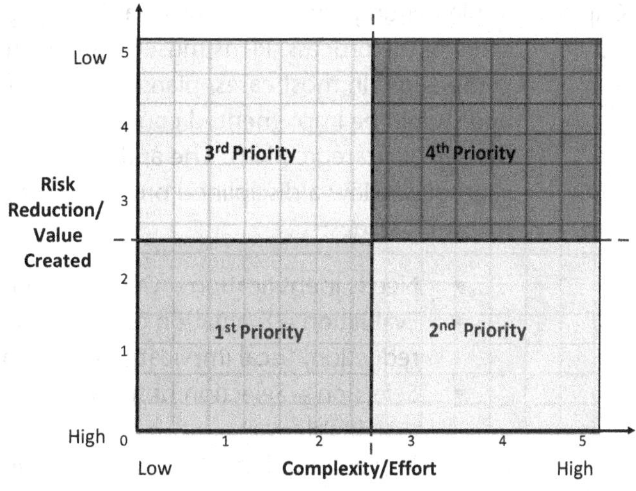

Source: © Safety Erudite Inc. (2019).

Figure 2.5: Model for prioritizing gaps for closure.

Developing Gap Closure Plans and Strategies

The best approach to developing gap closure plans involves leveraging the capacity and capabilities of networks. Through the collective efforts of the element network, members will generally do the following:

- Gather internal knowledge from all business units across the organization.
- Where necessary, seek knowledge and information from external sources to enhance internal processes.
- Evaluate and assimilate multiple sets of information into a single process (standardized) or consistent processes for the various business units.
- Develop simple, easy-to-apply solutions fit for purpose.

Figure 2.6 provides an overview of the process for developing gap closure plans and strategies.

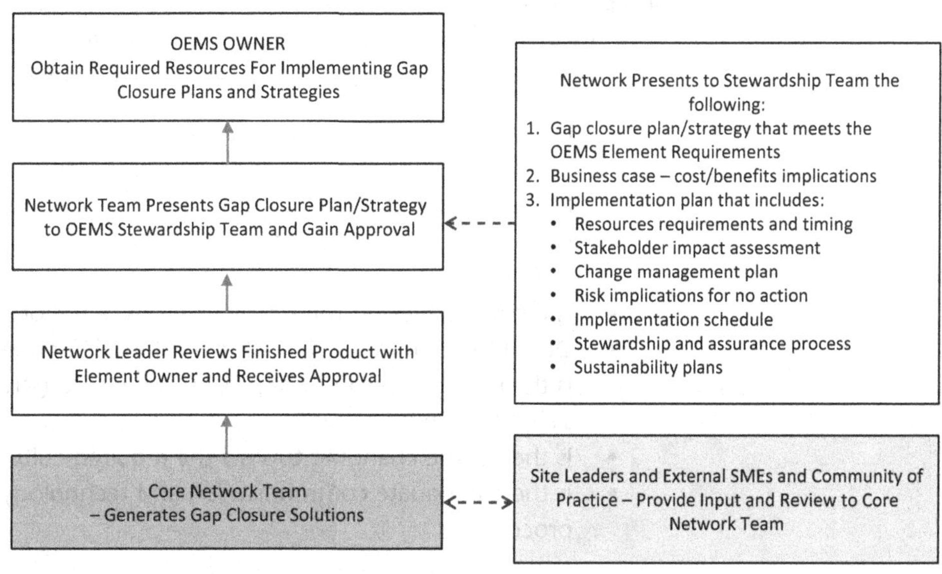

Source: © Safety Erudite Inc. (2019).

Figure 2.6: Process for closing gaps identified.

Implementing Gap Closure Plans and Strategies

Implementing gap closure plans and strategies is perhaps the most difficult stage in the process. Plans must be adequately resourced and stewarded to completion. In most cases, plans may fall into either a tactical or strategic basket and should be implemented consistent with the guidelines for implementing tactical or strategic plans. The authors recommend all gap closure plans and strategies follow a disciplined project management process including the following stages:

- Needs identification – Generally determined from the gap assessment
- Evaluation – Evaluation of alternative options that addresses risk reduction, fiscal implications, and value maximization
- Selection – Selection of the alternative that eliminates the gap in a sustainable way
- Implementation – Following a disciplined implementation process
- Closeout – Verification that the gap is closed in a sustainable way

2.9 Model 2 – Upgrading the Existing MS

Upgrading the MS – Overview

Upgrading the MS is often driven by the following:

- For HSE improvements – generally after a major incident
- As a proactive response to a major HSE incident in the industry
- For financial and operating performance improvements (including cost reduction)
- To reduce gaps between self and peers
- As a strategic initiative intended to transform the organization

(Ernst and Young, 2017)

Analysis and Reviews of Existing MS

In many instances, organizations may adopt an approach similar to that of Model 1 (Refer to Section 2.2). This is particularly so where the MS is fledgling and immature. However, once OEMS has been implemented for some time, 18–24 months, reviews and analyses are performed to determine the effectiveness of its components by answering the following questions:

- Is the scope and depth of each element (requirements and expectations) adequate for the element and business needs?
- Is the number of elements right for the organization?
- Is leadership fully engaged and onboard with the MS goals and objectives?
- Is the MS achieving the strategic intent of the organization?
- Does the organization have the right implementation process in place?
- Is the people aspect of the process addressed (structure and governance, skills and competence)?
- Is the culture changing toward the required culture?
- Is there adequate communication and technology to support the process?

Model for Analysis and Corrective Actions Management

A model for performing the analysis and review and proposed corrective actions management is provided in Figure 2.7.

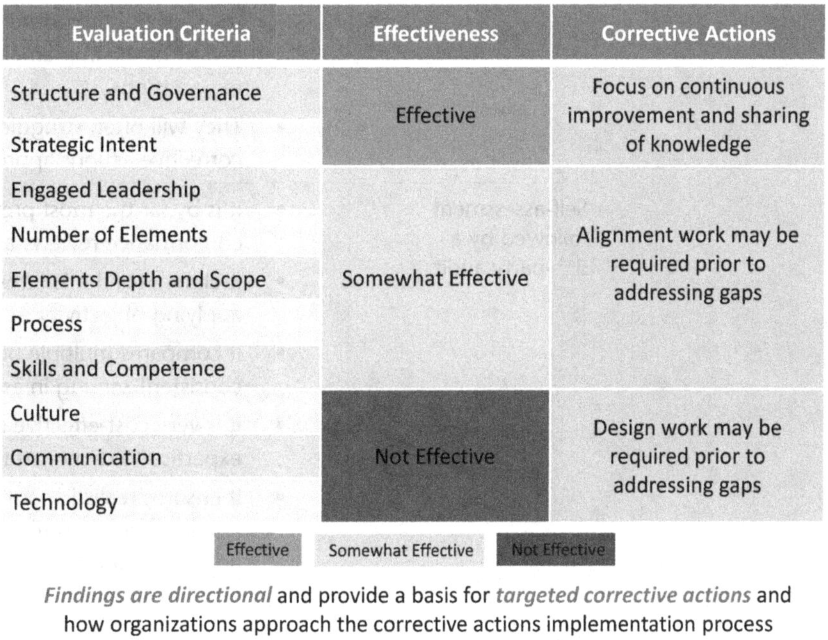

Evaluation Criteria	Effectiveness	Corrective Actions
Structure and Governance	Effective	Focus on continuous improvement and sharing of knowledge
Strategic Intent		
Engaged Leadership	Somewhat Effective	Alignment work may be required prior to addressing gaps
Number of Elements		
Elements Depth and Scope		
Process		
Skills and Competence		
Culture	Not Effective	Design work may be required prior to addressing gaps
Communication		
Technology		

Effective | Somewhat Effective | Not Effective

Findings are directional and provide a basis for *targeted corrective actions* and how organizations approach the corrective actions implementation process

Source: © Safety Erudite Inc. (2017).

Figure 2.7: Model for management system review and corrective actions management.

2.10 Conduct OEMS Conformance Reviews and Audits

Conformance Reviews and Audits

In today's business environment, most organizations adopt a series of conformance reviews and audits as shown in the following table:

Review/Audit Type	What We Mean
Self-assessment	• Where multiple business units are available, SMEs within the business unit undertake the review, comparing performance to requirements and expectations. • It is often a subjective review with a fairly accurate status of requirement implementation. • Failure of self-assessment is that some SMEs may be overly generous or conservative in assessment. • Self-assessments may require an independent review to ensure objectivity.
2nd -party audits	• Second-party audits are conducted across business units by a corporate audit team. • They are very objective in their reviews and audits. • They are costly relative to self-assessments.

3rd -party audits	• Third-party audits are performed on behalf of the organization by external service providers. • They will generally produce an audit report reflecting the views of those interviewed versus a detailed review of the implementation of requirements and expectations. • They will often struggle to generate acceptable corrective actions appropriate for the organization.
Self-assessment followed by a 2nd -party audit	• It is by far the most preferred approach to OEMS conformance reviews and audits. • It eliminates the failures of self-assessments by applying objectivity. • It compares multiple business units to ensure consistent scoring in assessments. • It is very cost-effective and leverages resident expertise to demonstrate conformance. • It ensures realistic corrective actions are developed and implemented to address gaps in conformance and compliance where applicable.

OEMS Maturity Scoring

Reviews and audits are conducted with the goal of assessing the maturity of the organization and allocate a score for objectivity. Scoring helps identify the maturity stage of the organization. Figure 2.8 provides an overview of scoring and the maturity stage of the organization.

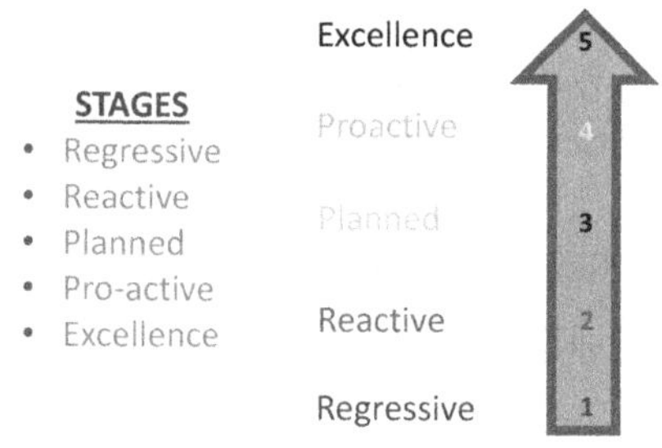

Source: © Safety Erudite Inc. (2019).

Figure 2.8: Overview of scoring associated with OEMS maturity.

Figure 2.9 provides an overview of the criteria used to determine the maturity of an element or of the MS as a whole.

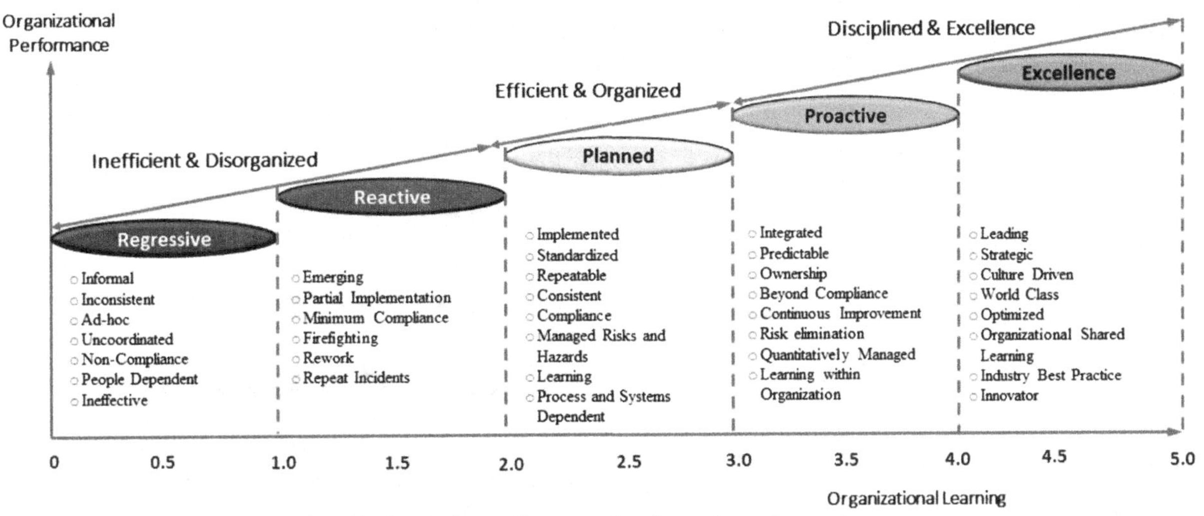

- Applicable Sources defined by level of compliance to the element requirements
- At the very minimum, the Organization must strive for planned compliance levels

Source: © Safety Erudite Inc. (2019).
Figure 2.9: Management system maturity map.

OEMS Maturity Map

The OEMS maturity map is a model that allows the organization to determine the following:

- Maturity of each OEMS element across the organization
- Overall OEMS maturity across the organization
- OEMS maturity within a business unit
- Comparative assessment of facilities within a business unit or business unit comparison

Figure 2.10 demonstrates how the criteria for scoring are applied to an element.

- Review the status of each requirement / sub-element relative to the measures that include Leadership Commitment; How well the risk exposures are being managed; Are the desired performance levels being derived; How well the requirement is being implement
- Assess against the Corporate OEMS Maturity Map
- Allocate score accordingly
- Identify gap/constraints in achieving a score of 3.0 on the OEMS Maturity Map

Element	*Sub –Element/Requirements*	*Gaps Identified*	*Avg. Score*
Management of Change	• A formal MOC Process must exist for managing change ○ A MOC tool is available and implemented across all Sites/Facilities ○ Requirement ***n*** • Change is authorized • A stakeholder impact assessment is completed and risks mitigated • Change is communicated • Changes is closed out • Sub-Element ***n***	• Standardized tools for managing change not available • Individuals identified vs. roles for assigned responsibilities • Inconsistent signoff of all required authorities for approving change • Very effective use of email, alerts, communication and training tools for communicating change • Inconsistent approached to managing the effectiveness of change. Documentation absent	2.0 2.0 2.5 3.5 1.5
Element Score			**2.3**

Source: © Safety Erudite Inc. (2019).
Figure 2.10: Mapping the maturity of an element.

Figure 2.11 demonstrates the maturity mapping of an element.

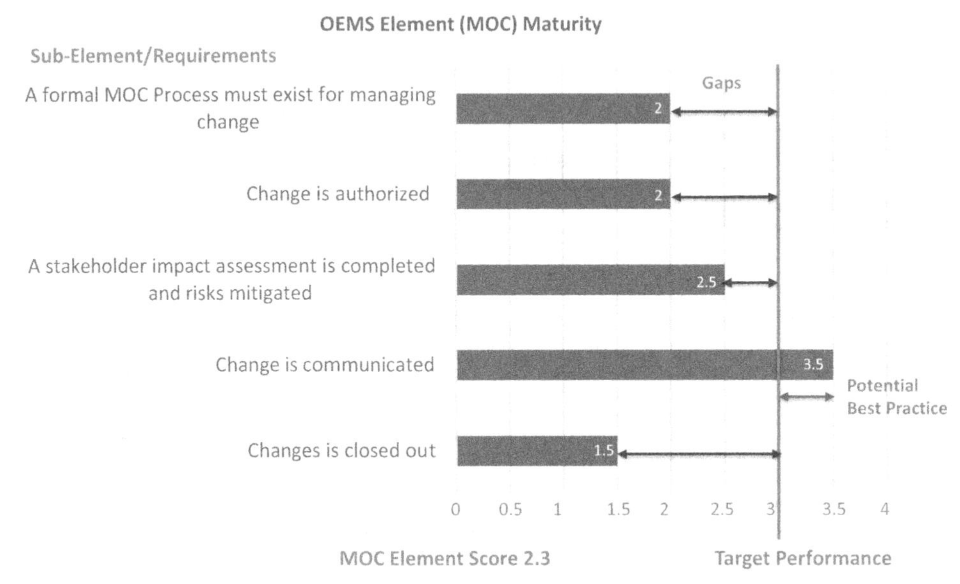

Source: © Safety Erudite Inc. (2019).
Figure 2.11: OEMS element maturity map.

Figure 2.12 demonstrates the maturity mapping of the corporate or a business unit OEMS maturity map.

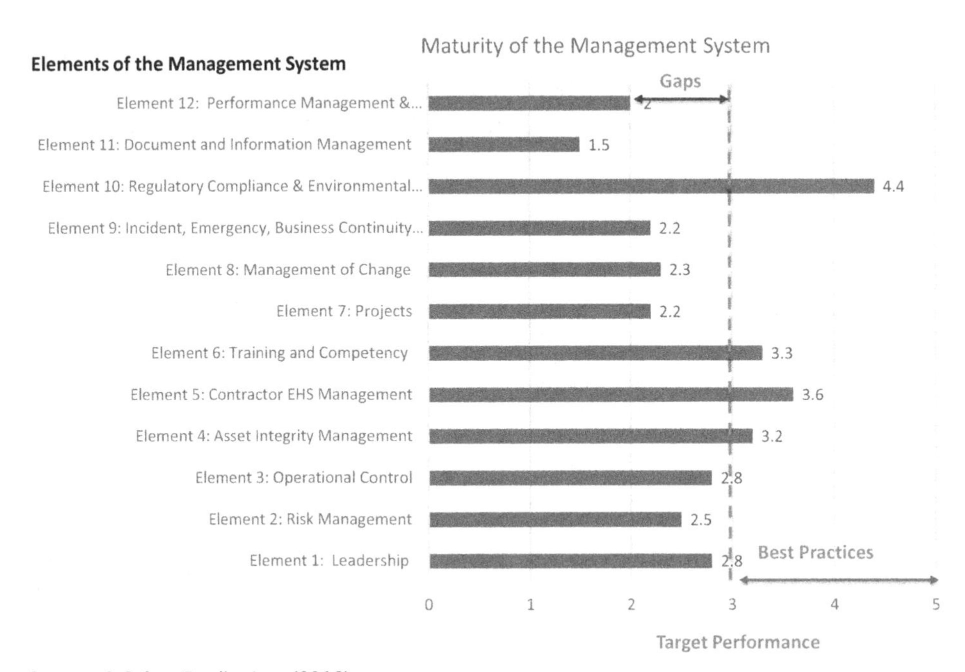

Source: © Safety Erudite Inc. (2019).
Figure 2.12: Corporate or business unit OEMS maturity map.

Overlay of multiple maturity maps developed for a business unit or facility allows for comparative assessment to determine opportunities for collaboration and sharing of best practices.

2.11 Identifying and Sharing Best Practices

Identification and Sharing of Best Practices

Best practices are identified from the review and audit processes that identify performance within the range of 3.0–5.0.

- Best practices provide opportunities for rapid transfer of knowledge and practices across the entire organization from business unit to business unit or from facility to facility within the business unit.
- Sharing of best practices may involve full-scale transfer of processes, control methods, technical knowhow, and people for implementation across business units.

2.12 Management Review of OEMS

Management Reviews

Management reviews of OEMS are required to assess the overall performance of the MS.

- They may be conducted on a 1-3-year frequency
- They are designed to determine the effectiveness of the MS and to take timely corrective actions
- They leverage opportunities generated from the MS and close gaps generated
- They are usually conducted by internal reviewers, but organizations may consider 3rd -party external reviewers

---

Corrective Actions Management

Management reviews may generate opportunities for improving the overall MS performance. Corrective actions may be generated that

- Close gaps identified and share best practices.
- Are assigned, resourced, and stewarded to completion.
- Are reassessed at some time in the future for sustainment.

2.13 Continuous Improvements and Sustainment

Continuous Improvement

Once element networks have completed their work in implementing the requirements and expectations of their respective element, networks transition to core responsibilities for continuous improvement.

Continuous improvement focuses on the following:

- Review of internal processes, tools, and practices and working effectively to deliver performance from implemented requirements and expectations
- Scanning the external business environment for more effective processes to achieve desired performance
- Functioning within external networks as SMEs to collaborate and share industry knowledge to improve business processes and practices

---

Sustainment

Once the requirements and expectations of an element have been implemented, organizational changes may test the resilience of the process. Sustainment of the element requires genuine organizational efforts in the areas identified in the following table:

Effort Focus	What We Mean
Governance	• Documenting and maintaining current standards, procedures, tools, and practices • Maintaining a trained and competent workforce • Providing adequate resources for maintaining relevant processes • Ensuring continuity when personnel changes are made
Best practices sharing	• Leveraging the full capabilities of element networks to ensure cutting-edge practices, processes, and technologies
Self-assessments	• Continuous review of the effectiveness of existing processes relative to the following: ○ Leadership commitment ○ Risk management ○ Performance expectation and process implementation
2nd or 3rd party reviews and audits	• Conducting audits of the effectiveness of existing processes relative to the following: ○ Leadership commitment ○ Risk management ○ Performance expectation and process implementation

References

Ernst and Young (2017). Driving Operational Performance in Oil and Gas. Retrieved August 03, 2017 from www.ey.com/gl/en/industries/oil---gas/ey-driving-operational-performance-in-oil-and-gas-1-low-oil-price-highlights-the-need-for-operational-excellence.

Lutchman, C., Evans, D., Maharaj, R., & Sharma, R. (2013). *Safety Management: Leveraging Networks and Communities of Practice for Continuous Improvements*, CRC Press.

Safety Erudite Inc. (2019). The Integrated Process Management System Provider. Retrieved June 2019 from www.safetyerudite.com

3 READYING THE ORGANIZATION FOR OEMS

Overview

Readying the organization for OEMS is a critical step in the process for successful and sustainable implementation of an OEMS. This process requires a series of actions and work essential for creating the momentum for change. Key activities include the following:

- Creating a shared OEMS vision
- Securing leadership commitment
- Engaging and involving key stakeholders
- Establishing the roles and responsibilities of key OEMS stakeholders
- Building the OEMS organization
- Developing the OEMS framework
- Communication – Seven times, seven different ways

In this chapter, the authors detail the process for creating the momentum for change and for readying the organization for change.

3.1 Creating a Shared OEMS Vision

What Is Commitment?

Commitment to organizational change is defined as "a force (mind-set) that binds an individual to a course of action deemed necessary for the successful implementation of a change initiative" (Herscovitch and Meyer, 2002, p. 475; Hill, Seo, Kang, and Taylor, 2012). They advised employee commitment to change can be categorized as follows:

Commitment	Definition
Affective	Affective commitment to change (ACC) is a "desire to provide support for the change based on a belief in its inherent benefits" (Hill et al., 2012, p. 758)
Normative	Normative commitment to change (NCC) is a "sense of obligation to support the change because it is one's duty to do so" (Hill et al., 2012, p. 758)

Hill et al. (2012) also advised that building commitment to major or radical change is the responsibility of organizational leaders. This is particularly so for OEMS, since in most instances OEMS is regarded as a radical change and is often initiated by senior leaders of the organization.

Commitment – Maslow's Hierarchy Revisited

Stum (2001) in a revisit of Maslow's hierarchy of needs suggests "there is a hierarchy of organizational factors that build upon one another to construct higher levels of commitment in the workforce" (p. 9). Stum (2001) also proposed a workforce commitment model as shown in Figure 3.1.

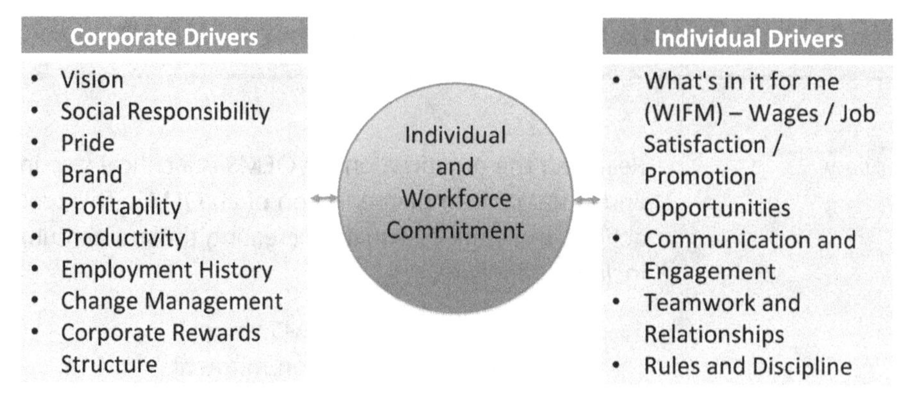

Source: Stum (2001) – Maslow revisited: Building the employee commitment pyramid.
Figure 3.1: Workforce commitment model.

Commitment to OEMS at all levels of the organization is determined by how workers perceive improvements for themselves in the performance pyramid as a result of OEMS implementation and development across the corporation.

Creating the OEMS Vision

Readying the organization for OEMS is a critical step in the process for successful and sustainable implementation of an OEMS.

- The vision should reflect the progress of the organization toward excellence and highlight a workplace characterized by excellence in environmental, health, and safety (EH&S), reliability, and business performance
- The OEMS vision must be realistic and achievable, and should create a compelling argument for followers to be a part of the vision
- The vision should resonate with people and provide answers to the stakeholder question: *What's in it for me* (*WIFM*)

Communicating the OEMS Vision

Communicating the OEMS Vision is a joint exercise between the highest authority of the organization – president/CEO and other senior business leaders such as senior vice presidents (SVPs) and executive vice presidents (EVPs).

- The OEMS vision must be documented in the OEMS framework brochure that is accessible to all employees of the organization.
- Senior leaders should seize every moment to share the OEMS vision in person with all levels of the organization.
- Communicating the vision requires structured and consistent messages.
- The vision may be communicated using the past, current, and future state as shown in Table 3.1.

Table 3.1: Past, Current, and Future State Messages for Establishing the OEMS Vision

Status 5 Years Ago	Status Today	Where We Want to Be 5 Years from Now
• Multiple assets operating very independently with different cultures • High dependence on people, little structure • Reactive and inconsistent • Weak allocation of resources based on risk across the company • Auditors says you do some good stuff but nothing is documented • Repeat of the same or similar incidents • Good efforts to improve processes that eventually went away • Resistance to sharing – not invented here • Duplicated efforts and rework • Inadequate or inconsistent setting of expectations • Lack of empowerment – requirements and expectations not documented and changes with the leader	• Great progress toward a consistent and uniform culture • Highlight the good things that are happing in the organization • Highlight the bad things that need to be fixed	• Multiple assets all operate safely with common purpose, culture, and pride • People matter, but we have the structure to carry on when they leave • Consistent and proactive • Thoughtful and strategic risk management and resources allocation across the company • Auditors audit us against our standards and procedures • We learn from incidents when we have them and share with each other • Processes are consistent where they need to be and customized where they need to be, but no matter what we sustain and improve them • We demand sharing and collaboration • We leverage our scarce resources – divide and conquer • Mature networks – the heavy lifting will be done, we will be sustaining and improving • Everyone knows what is required of them to run safely and reliably • We empower our employees because we have clearly defined roles and set boundaries • We have defined their decision rights and never second guess

3.2 Securing Leadership and Employee Commitment

Overview

Securing leadership and employee commitment to OEMS is critical. Leadership commitment is essential since respective elements of the management system are owned by senior leaders of the organization. Table 3.2 provides a typical ownership structure of OEMS elements. While the table identifies only a few senior leaders of the organization, leadership commitment at all levels of the organization is required.

Table 3.2: Senior Leadership Ownership of OEMS Elements

Element No.	Owner	OEMS Element Title
1	President/CEO	Leadership and organizational effectiveness
2	VP Corporate EH&S	Hazard identification and risk management
3	VP Corporate EH&S	Working safely
4	VP Corporate EH&S	Environmental stewardship
5	VP Legal and General Counsel	Legal and regulatory compliance
6	VP Operations	Asset reliability and integrity
7	VP Operations	Conduct of operations
8	VP Corporate EH&S	Management of change
9	VP Human Resources	Training, competency, and human performance
10	VP Procurement	Procurement – contractor and supplier management
11	VP Corporate EH&S	Emergency preparedness, security, and business continuity
12	VP Information Technology (IT)	Document and information management

Source: © Safety Erudite Inc. (2019).

Commitment at all levels of the organization is achieved in the following ways:

- Sharing the corporate and OEMS vision
- Ensuring a clear line of sight among tactical plans, strategic plans, and the corporate vision
- Prevalence of strong transformational leadership behaviors
- Proactive actions to
    - Reorganize personnel when applicable to ensure good fit
    - Remove personnel from the organization who are not committed and may hinder/derail the process

Building Commitment

Studies show that transformational leadership behaviors are positively related to change commitments among followers. Behaviors demonstrated by transformational leaders include:

- Create a shared vision; promotes involvement, consultation, and participation
- Encourage creativity, innovation, proactivity, responsibility, and excellence
- Moral authority derived from trustworthiness, competence, sense of fairness, sincerity of purpose, and personality
- Lead through periods of challenges, ambiguity, and intense competition or high growth periods
- Promote intellectual stimulation and considers the individual capabilities of each employee
- Are willing to take risks and generate and manage change
- Lead across cultures and international borders
- Build strong teams while focusing on macro-management
- Are charismatic and motivate workers to strong performance

Herold, Fedor, Caldwell and Liu (2008) suggested that transformational leadership behaviors are important during change because of the "ability of transformational leaders to engage followers and motivate them to support the leader's chosen direction" (p. 353). Bass and Avolio (1997) similarly advised transformational leaders motivate employees by inspiring hearts and minds (emotional appeal) and by creating a shared, compelling vision. In their view, transformational leaders encourage employees to venture beyond personal interests and to work together in the best interests of the collective (e.g., team, organization). Timothy, John and Xiao-Hua (2013) suggested that organizations should seek to encourage transformational/charismatic leadership within organizations as a means of building affective commitment (where employees want to stay), regardless of the cultural context of the organization.

It is imperative, therefore, that OEMS be supported with strong transformational leaders who inspire the hearts and minds of the organization's leadership team. Moreover, such behaviors should infiltrate the entire organization to inspire workers and build commitment throughout the organization to the OEMS vision.

Leadership Behaviors That Undermines Commitment

Leaders must be vigilant in identifying behaviors that undermine commitment to OEMS. Table 3.3 lists some of the undesirable leadership behaviors that can adversely impact OEMS implementation. These behaviors must be adequately addressed by leaders through timely intervention and corrective actions.

Table 3.3: Leadership Behaviors Affecting Commitment

Leadership Behaviors	Details
Cynics	• Cynicism is a psychological disposition by leaders that distrusts everything or selectively trusts only what he/she knows. • A cynic has a lack of faith in everyone. He/she elevates distrust in others and trust only in self.
Skeptics	• Skeptics refer to those who inject so much doubt that the uncertainties of risk and human fallibility become a block to the OEMS process. • Skeptics generally do not know what to do except they know it is not what you are doing.

Pessimist	• With pessimism and negativity everything tends to just spiral downward. • A pessimist is fueled by skepticism and cynicism that leads to a very unhappy workplace that pulls everyone down.
Double speak	• Gap between what we say and what we do. • When this gap is at extremes in contradiction, employees soon learn not to trust the words of others.

Source: © Safety Erudite Inc. (2019).

3.3 Engaging and Involving Key Stakeholders

Overview

Engaging and involving key stakeholders in the OEMS process is a critical component of the entire process and for building commitment. When people are engaged and involved, ownership and commitment tend to be high. With OEMS, the goal is to create ownership. As shown earlier in Figure 2.8 (Criteria Used to Allocate Maturity Score), ownership is a criterion used in determining the OEMS maturity level of the organization.

In this book, engagement and involvement are designed to represent different things and are not used interchangeably. Our definitions are as follows:

	Definition
Engagement	Engagement refers to the following: • Communication and providing adequate understanding to the stakeholder of a concept or process • Peripheral input provided by the stakeholder groups • Input may be accepted or rejected in the final product
Involvement	Involvement refers to the following: • Active stakeholder participation in developing a concept or process • Provides core input into the final product • Involves effective owners who are responsible for selling the core messages of the produce/process

Who Are Engaged?

In the OEMS development process, the goal is to engage all workers and internal stakeholders of the organization. Employees and contractors at all levels of the organization are engaged at varying extents. Engagement may include the following key stakeholder groups:

Stakeholder Groups	Examples of Engagement

Workers at the front line (includes operators/ mechanics and trade persons)	Engagement of frontline workers generally includes the following:
	• A principal goal of providing adequate understanding of the proposed change and how the change will impact them
	• Generally, a communication process involving the use of various tools and processes
	  ○ May require group communication with senior respected leaders of the organization who understand OEMS, its goals, and impact to the stakeholder group very well
	  ○ Will require periodic follow-up engagement sessions in the form of "question and answers"
	  ○ May use brochures, presentations, print materials, and online messaging
	• Engagement is important since this group will be impacted with a more disciplined approach to doing work

Frontline supervisors	Engagement of frontline supervisors generally includes the following: • A principal goal of providing adequate preparation for the proposed change and how the change improves business performance • Will include training and development in the new/ improved ways of doing things
Middle managers (includes department leaders)	Engagement of middle managers generally includes the following: • A principal goal of seeking knowledge on how to improve operating processes • Inputs in defining minimum requirements and expectations of the business
Functional groups and organizations	Engagement of functional groups and organizations generally includes the following: • A principal goal of providing adequate understanding to the stakeholder of the proposed change and how the change will impact the support required from the functional groups • Some functional leaders may provide subject matter expertise and will therefore require a more in-depth understanding of OEMS and the respective element stewarded by the function

Who Are Involved?

In the OEMS development process, people are involved to promote ownership and to ensure the right requirements and expectations of the organization are documented, defined, and communicated to stakeholder groups. Involved stakeholders may include the following:

Stakeholder Group	Examples of Involvement
President and/or CEO	Involvement of the President and/or CEO refers to the following: • Creating a compelling argument for OEMS • Confirming and articulating the OEMS vision of the organization to senior leaders • Identifying key senior leaders and stakeholders with responsibilities and accountabilities relating to OEMS • Ensure adequate resources are available
Senior organizational leaders	Involvement of senior organizational leaders refers to the following: • Assigning authority and responsibilities for coordinating and stewarding OEMS • Resourcing all related work • Reviewing and approving organizational requirements and expectations
OEMS element owners	Involvement of OEMS element owners refers to the following: • Assuming authority and responsibilities for coordinating and stewarding owned OEMS elements • Communicating the requirements and expectations for owned elements across the organization • Identifying network members responsible for providing subject matter expert (SME) contributions and site/facility ownership, responsibilities, and accountabilities
Network members	Involvement of network members refers to the following: • Assuming site/facility-specific authority and responsibilities for implementing and stewarding owned OEMS elements requirements at the site/facility • Communicating site/facility-specific requirements and expectations for owned elements • Providing SME contributions regarding assigned elements
SMEs and consultants	Involvement of SMEs and consultants refers to the following: • Providing input into developing requirements and expectations • Providing expert contributions to implementation processes and continuous improvements

3.4 Establishing the Roles and Responsibilities of Key OEMS Stakeholders

Identifying Key OEMS Stakeholders

Preparing the organization for OEMS requires clearly defined roles and responsibilities for involved stakeholders. The President, CEO, or delegate is responsible for identifying key OEMS stakeholders and defining roles and responsibilities among these stakeholders. Typical OEMS roles include the following:

- President and/or CEO
- VP Operations Excellence
- Operations Excellence Stewardship team
- OEMS element owner
- OEMS coordinator
- OEMS network member
- Technical network member
- SMEs and consultants
- Functional organization leaders
- All workers (including contractors)

Stakeholder Roles and Responsibilities

Roles and responsibilities of personnel involved in OEMS must be clearly defined and understood by the respective stakeholder. Key stakeholder roles and responsibilities are shown in the following table:

Role	Responsibilities
President and/ or CEO	Responsibilities of the President and/or CEO include the following: • Confirming and articulating the OEMS vision of the organization to senior leaders • Making OEMS the central focus within the organization • Resourcing OEMS across the organization • Supporting the OEMS VP in the implementation of OEMS
VP Operations Excellence	Responsibilities of the VP operations excellence include the following: • Sharing the OEMS vision throughout the organization • Positioning OEMS as a central focus of the way business is done in the organization • Building the OEMS organization – having the right person in the right role • Stewarding the implementation of OEMS – prioritizing and resourcing OEMS work across the organization • For multiple sites and facilities, supporting the OEMS coordinators in the implementation of OEMS requirements and expectations • Enabling the process by removing obstacles and barriers to the adoption of OEMS requirements and expectations

Operations Excellence Stewardship team	Responsibilities of Operations Excellence Stewardship team include the following: • Reviewing overall progress on the implementation of OEMS across the organization • Evaluating business cases for prioritized OEMS implementation activities • Establishing the schedule for OEMS audits – 2nd - and 3rd party audits • Assessing the cultural shift of the organization toward operations excellence
OEMS element owners	Responsibilities of OEMS element owners include the following: • Ensuring requirements and expectations of the element are appropriate and reflective of the desired state • Identifying OEMS element network members responsible for implementing element requirements and expectations at respective sites/facilities • Identifying technical network members responsible for providing SME contributions for continuous improvements • Communicating the requirements and expectations for owned elements across the organization at the request of the OEMS VP or coordinator • Providing input in prioritizing OEMS implementation work based on risk exposure and value creation for the organization
Site/facility VP or leader	Responsibilities of site/facility VP or leader include the following: • Assisting in prioritizing OEMS work for the site/facility • Resource-prioritized work • Communicating to site/facility personnel the goals and objectives of OEMS, progress updates, and priorities • Creating a culture of operations discipline for the site/facility
OEMS coordinator	Generally applicable where multiple sites/facilities are available, responsibilities of OEMS coordinators include the following: • Coordinating the implementation of OEMS for assigned sites/facilities • Providing consistent interpretation of OEMS requirements and expectations across the site/facility • Becoming the go-to person for all OEMS requirements, expectations, and activities • Eliminating barriers and challenges to the VP OEMS where applicable • Generating OEMS implementation progress reports and presenting to the OEMS stewardship team • Supporting the work of site/facility OEMS and technical network members • Guiding the OEMS network in prioritizing work and developing the business case for prioritization and implementation

OEMS network members	Responsibilities of network members include the following: • Becoming the site/facility expert for the assigned OEMS element • Implementation and sustainment of assigned OEMS elements at the site/facility • Communicating site/facility-specific requirements and expectations for owned elements • Providing SME contributions at the site/facility regarding assigned elements
Technical network members	Technical networks are formed to deliver excellence in specific areas related to an OEMS element. When formed, responsibilities of technical network members include the following: • Providing SME guidance and input into specific requirements and expectations of an OEMS element • Providing expert contributions to implementation processes and continuous improvements for the specific requirement and expectation
SMEs and consultants	Responsibilities of SMEs include the following: • Providing input specific to OEMS and technical requirements and expectations • Providing expert contributions to implementation processes and continuous improvements
Functional organization leaders	Responsibilities of functional leaders include the following: • When assigned an OEMS element, they are responsible and accountable per the OEMS element owner • Providing support and guidance to the OEMS coordinator and VP OEMS in implementation processes as required • Providing knowledge and expertise for continuous improvements
All workers (including contractors)	Responsibilities of all workers (including contractors) include the following: • Performing all work safely and consistent with the defined requirements and expectations of the organization • Following procedures when performing work • Be trained and competent for assigned work • Reporting all unsafe conditions and actions • Refusing all unsafe work • Providing expert contributions to implementation processes and continuous improvements

3.5 Building the OEMS Organization

Stewardship Organization

Having the right organization in place is essential for effective implementation and stewardship of OEMS. Three distinct teams within the OEMS organizations are required for success in implementation and sustainment. They are as follows:

- The Implementation Team
- The Enablement Team
- The Stewardship Team

Figure 3.2 provides an overview of the various team compositions and the relationships within the OEMS organization.

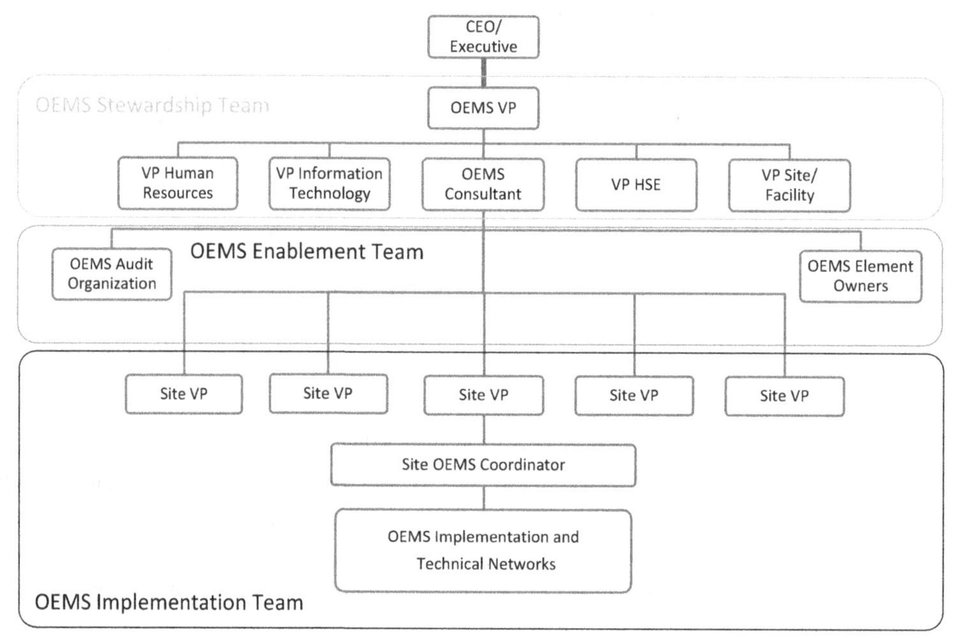

Figure 3.2: Typical OEMS organization and associated teams.

OEMS Element Networks

OEMS element networks are developed with input from the site/facility VP or leader. These networks are designed to lead the implementation and sustainment of the element requirements and expectations for the site/facility.

- OEMS element network leaders report to the OEMS coordinator for each site/facility.
- Network members collect knowledge relating to the element from all sites/facilities. This knowledge is analyzed and used to develop the most effective processes for all sites/facilities.
- Each network has a network leader.
- Networks are chartered with fixed annual deliverables which become part of each member's work plan.
- Network members are asked to do more than their normal workloads without the expectation of additional remunerations.
- Excellence in performance may be rewarded through a President's Award for operations excellence.

The design of OEMS element networks is shown in Figure 3.3.

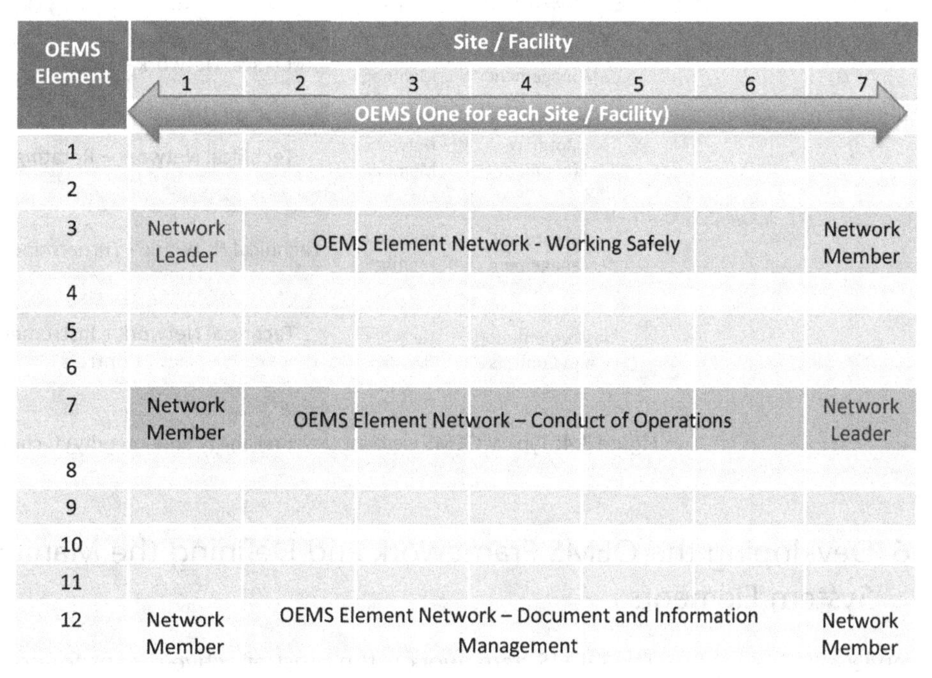

Figure 3.3: Typical OEMS element networks.

OEMS Element Technical Networks

OEMS element technical networks are developed with input from the site/facility VP or leader and departmental leaders. These networks are designed to provide expert knowledge of how to deliver on OEMS element-specific requirements and expectations for the site/facility.

- Each network has a network leader. OEMS element technical network leaders may report to the OEMS element coordinator or the discipline leader for the site/facility.
- Technical network members collect knowledge relating to the requirement and expectation from all sites/facilities. This knowledge is analyzed and used to develop the most effective processes for all sites/ facilities.
- Networks are chartered with fixed annual deliverables which become part of each member's work plan.
- Network members are asked to do more than their normal workloads without the expectation of additional remunerations.
- Excellence in performance may be rewarded through a President's Award for operations excellence.

The design of OEMS element networks technical network is shown in Figure 3.4.

Element 6 - Technical Networks	Site / Facility						
	1	2	3	4	5	6	7
Electricals Management	Network Member	Technical Network – Electricals Management					Network Member
Rotating Equipment	Network Member	Technical Network – Rotating Equipment					Network Leader
Turnaround Management	Network Member	Technical Network – Turnaround Management					Network Member
Instrumentation and Controls	Network Leader	Technical Network – Instrumentation and Controls					Network Member

Figure 3.4: Typical OEMS element (asset reliability and integrity) technical networks.

3.6 Developing the OEMS Framework and Defining the Management System Elements

Overview

The OEMS framework is the most effective means for communicating the high-level, directional vision of the organization. Careful attention is required to developing the OEMS framework to ensure the following:

- Easy access across all levels of the organization
- Easy-to-read and unintimidating
- Written in simple language
- Realistic requirements and expectations
- Non-overwhelming
- Leadership commitment is clearly communicated
- Identifies the OEMS elements (9–12) and ownership
- Simple, yet impactful in communicating OEMS path and vision

Leveraging Industry Knowledge

Building the OEMS framework should leverage industry-shared and publicly available information and knowledge on peer and competitor management systems. Sources of shared and publicly available information may include the following:

- Company websites
- Industry associations (International Association of Oil and Gas Producers (IOGP) and International Association of Standardization (ISO))
- Reaching out to model organizations for guidance and help
- Collaboration within industry networks
- Industry conferences and forums
- University libraries
- Corporate annual reports

How to Build the Framework and Define the Elements

Building the OEMS framework requires input from senior leaders, communication experts, and SMEs. The following steps are required in developing the OEMS framework:

1. Gather management system framework brochures/magazines and documents from similar organizations.
    a. Collect from more than five selected industry-leading organizations.
2. Lay out all elements and sub-elements identified from the organizations selected.
3. Identify elements and sub-elements common to your business.
4. Repackage elements into 9–12 elements.
    a. Trend toward fewer elements for smaller organizations.
    b. Fewer elements reduce the management burden.
    c. Engage and involve senior leaders to right-size for your organization business needs and vision.
5. Identify and select relevant information communicated by other organizations that may be valuable to your efforts.
6. Create draft versions and review through several iterations with the OEMS VP, element owners, and President/CEO for finalization.
7. Finalize with President/CEO approval and be ready for communication sessions across the organization.

Layout of the Framework

There are many layouts possible for the OEMS framework. The one that is most appropriate, however, should communicate the core attributes of OEMS to all levels of the organization with the least effort. The authors recommend the following layout for the OEMS framework:

Section	Content

Introduction	The introductory section should include the following:
	• A table of contents
	• A joint introductory letter from the President/CEO and the OEMS VP communicating the OEMS vision and why OEMS is required
	• The OEMS vision
	• How OEMS builds upon the corporate values
	• Leadership commitment – feature prominent corporate leaders
	• Explanation of why OEMS is important to the organization and the workplace
	• Highlight the need for everyone to be a part of the process
	• Elements and their owners
	• Implementation overview
	  ○ Start date
	  ○ Implementation strategy
	  ○ Milestone events
	  ○ Highlight a 3- to 5-year implementation period

Elements and sub-elements	The elements and sub-elements section should include the following: • Elements numbered based on the prioritization of the organization – generally leadership and organizational effectiveness or variants first • A summarized objective of the element • Sub-elements (core focus) and requirements and expectations • Existing tools and processes that help meet requirements as applicable
Graphics, images, and key messages	Graphics, images, and key messages should include the following: • Graphics of simple and existing and tools processes that enables OEMS. For example, graphics of the corporate incident management tool, the corporate risk matrix, and procurement processes • Selected images of key personnel across all levels of the organization who are advocates of OEMS with key value statements or quotations • Key messages strategically placed to catch and retain attention. For example, operations discipline, life-saving principles, and walk the line.

3.7 Developing the OEMS Element Standards

Overview

Developing the OEMS element standard requires input from element owners and SMEs to ensure requirements and expectations are accurate and appropriate. Developing the element standards should be an in-house process leveraging input from senior leaders, departmental managers, network members, and frontline supervisors. Occasionally, organizations may place the development of these standards into the hands of external consulting organizations. External development of these standards will require extensive review – and rewriting in most instances – and can be quite costly.

Steps in Developing the OEMS Element Standards

When done internally, the following steps are required in developing the OEMS element standards:

1. Gather element standards from various sites/facilities as available.
   a. Collaborate with leading industry peers for sharing of sample standards.
   b. Explore Internet sources for sample standards
2. Structure element standards consistently with the sub-elements presented in the framework.
3. Detail requirements and expectations for each sub-element.
4. Write each element standard to reflect the OEMS vision for the organization – avoid documenting requirements and expectations reflecting only the current situation of the organization.
   a. Engage a single competent writer to document all element standards to ensure standardization as applicable, and consistency in documentation.

   b. Write for users – avoid writing for readers.
   c. Write for a Felsch–Kincaid reading grade level of 8–12.
5. Create draft versions and review with SMEs and element owners through several iterations for finalization.
6. Obtain President/CEO approval and be ready for communication sessions across the organization.

Figure 3.5 provides an overview of the process for preparing and developing the standard.

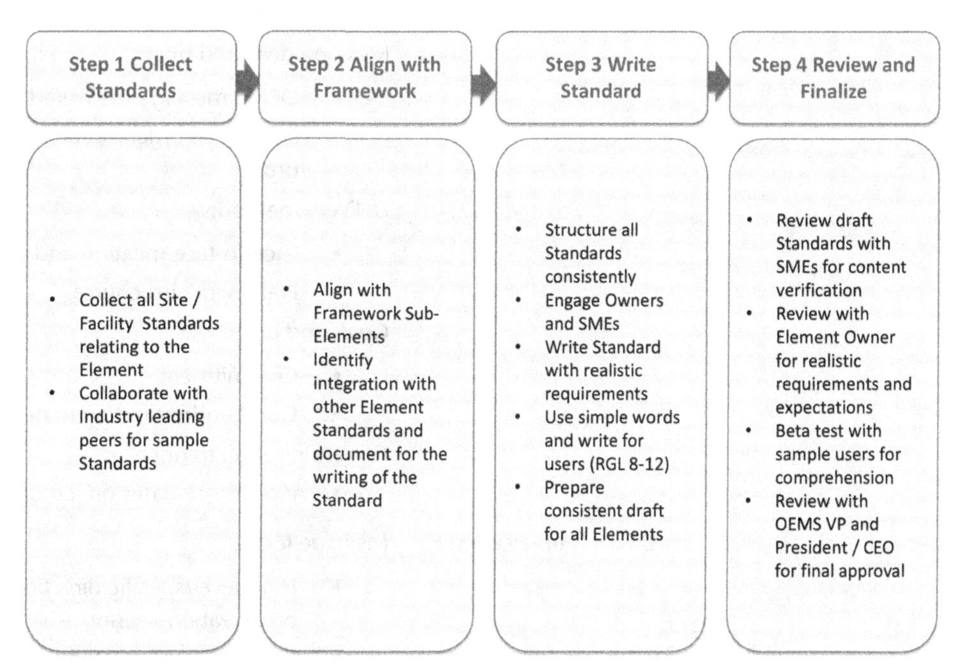

Source: © Safety Erudite Inc. (2019).
Figure 3.5: Steps in developing OEMS element standards.

3.8 Communicate … Communicate … Communicate

Overview

Effective and complete communication across all levels of the organization is perhaps the most important activity involved in readying the organization for an OEMS launch. Communication generally involves the use of multiple communication channels and should be designed for all levels of the organization. For effective communication, the following must be considered:

- The message being delivered
- The audience targeted
- Who delivers the messages
- The media used for delivering the message
- The goals and objectives of the communication
- The frequency of communication

The authors recommend communicating seven times seven different ways

Stakeholder Groups and Communication Considerations

The following table provides a summary of recommended communication methods and messages for OEMS stakeholder groups:

Stakeholder Groups	Communication Methods and Messaging
Senior leaders	Message focus: • High-level direction of the organization – vision • Business areas most impacted • Functional support required • Schedule and cost management Message delivered by: • OEMS messages delivered by President/CEO • Functional organization leader (VP) to functional group Delivery method: • Face-to-face meeting and presentations • Framework brochures Goals and objectives: • Commitment • Good understanding of business impact Communication frequency: • Progress updates on a monthly/quarterly basis
Business unit leaders	Message focus: • High-level/specific direction of the organization – vision • Business areas most impacted • High-priority focus areas for the business unit • Schedule and cost management Message delivered by: • Initial OEMS messages delivered by President/CEO • Subsequent messages delivered by OEMS VP and business unit leader Delivery method: • Face-to-face meeting and presentations • Framework brochures Goals and objectives: • Commitment • Good understanding of business impact Communication frequency: • Progress updates on a weekly/bi-weekly/monthly basis

Networks/OEMS coordinators and element owners	Message focus: • High-level and specific direction of the organization – vision • Critical need for teamwork, collaboration, and sharing • Network support required – roles and responsibilities and criticality of the network's role in the success of OEMS Message delivered by: • Initial OEMS messages delivered by President/CEO/ OEMS VP • OEMS VP and selected element owners Delivery method: • Network conference/workshop – all members hear the same message at the same time • Face-to-face meeting and presentations • Framework brochures Goals and objectives: • Commitment • Excellent understanding of business impact Communication frequency: • Progress updates on an ongoing weekly/bi-weekly monthly/quarterly basis
Department (middle) managers and frontline supervisors	Message focus: • High-level and specific direction of the organization – vision • Need for collaboration and sharing – collaboration versus competition • Functional support required • WIFM Message delivered by: • OEMS messages delivered by President/CEO • Element owners Delivery method: • Initial face-to-face meeting and presentations • Framework brochures Goals and objectives: • Commitment • Good understanding of business impact Communication frequency: • Progress updates on a monthly/quarterly basis

Frontline (all) workers	Message focus: • High-level and specific direction of the organization – vision • Need for operations discipline • WIFM Message delivered by: • OEMS messages delivered by President/CEO/OEMS VP/site/facility leader • Participation from department leaders and element owners Delivery method: • Initial face-to-face meeting and presentations (maximum 1 h) • Framework brochures • Intranet messages and videos Goals and objectives: • Commitment Communication frequency: • Progress updates on a monthly/quarterly basis

References

Bass, B. M., & Avolio, B. J. (1997). *Full Range Leadership Development: Manual for the Multifactor Leadership Questionnaire*. Palo Alto, CA: Mindgarden.

Herold, D. M., Fedor, D. B., Caldwell, S., & Liu, Y., (2008). The effects of transformational and change leadership on employees' commitment to a change: A multilevel study. *Journal of Applied Psychology, 93*(2), 346–357. Retrieved December 24, 2017 from EBSCOHost Database.

Hill, N. S., Seo, M., Kang, H., & Taylor, S. (2012). Building employee commitment to change across organizational levels: The influence of hierarchical distance and direct managers' transformational leadership. *Organization Science, 23*(3), 758–777. ISSN 1047-7039 (print) — ISSN 1526-5455 (online). Retrieved December 24, 2017 from EBSCOHost Database.

Safety Erudite Inc. (2019). The Integrated Process Management System Provider. Retrieved June 2019 from www.safetyerudite.com

Stum, D. L. (2001). Maslow revisited: Building the employee commitment pyramid. *Strategy & Leadership, 29*(4), 4–9. Retrieved December 24 from ProQuest Database.

Timothy, A. J., John, P. M., & Xiao-Hua, W., (2013). Leadership, commitment, and culture: A meta-analysis. *Journal of Leadership and Organizational Studies, 20*(1), 84–106. Retrieved December 24, 2017 from EBSCOHost Database.

4 REPACKAGING MULTIPLE MANAGEMENT SYSTEM INTO AN OEMS

Overview

Most organizations today are the outcome of the following business and growth strategies:

- Organic growth
- Mergers and acquisitions
- Joint ventures

As a consequence of these growth strategies, many of these organizations are comprised of multiple inherited management systems and cultures. Indeed, this was the case in BP that review suggests contributed to the deadly Texas City BP Incident in 2005 (U.S. Chemical Safety and Hazard Investigation Board, 2007). Moreover, legacy systems which evolved with organic growth are slow (or have failed) to respond adequately to advances in business processes as they seek profitability versus long-term sustainability that comes with a structured and documented management system.

The prevalence of multiple management systems is high from growth through mergers and acquisitions. In most instances, where organizations are comprised of multiple acquired assets from different mother organizations, the prevailing management system is inherited throughout the acquisition and turnover period. The challenge for most organizations with multiple management system is to transition to a single management system while minimizing the change impact and disruption to the business. In this section, the authors provide an overview of the process for transitioning from multiple management systems toward a single and fully integrated OEMS.

4.1 Consolidating Existing Management System under a Single Umbrella

The OEMS Vision

Brien and Meadows (2000) advised that most authors agree that the development of the corporate vision depends on a process of internal and external analysis ultimately leading to the vision statement. They also recommend a series of steps as follows:

1. Analyze company's future environment
2. Analyze future competition
3. Analyze company's resources and core competencies
4. Clarify organizational values
5. Develop vision statement
6. Contrast vision with the present state of the organization
7. Use vision to develop strategic objectives, goals and options (p. 37).

Llewellyn (2005) advised that in the health care industry, the CEO sets the vision of the organization to gain worker loyalty. Llewellyn (2005) also advised that a successful and sustainable vision is one that, in addition to business profitability measures, embraces "the human spirit of sensing the needs of others through meaning, purpose, empathy, caring, and sharing" (p. 374). Not unlike the health care industry, the OEMS vision must appeal to the hearts of all workers thereby encouraging them to share the vision, and driving them to become a part of the movement created by the shared OEMS vision.

Defining the OEMS Elements

Defining and packaging the elements of OEMS is primarily about what works best for the organization's stewardship process. The following table provides guidance in defining and packaging the OEMS elements:

Element Attribute	Guidance
Title	Titling considerations should include the following: • Include leadership commitment as a central theme or OEMS element • Align with OSHA process safety management (PSM) elements where possible • Provide simple titles that are easy to refer to by all levels of the organization • Ensure core functional organizations are represented – HR, procurement, legal, etc.
Numbering	Numbering considerations should include the following: • Number OEMS elements in a hierarchical manner with the most important element identified first • Start with leadership commitment • Place people-focused elements next – for example, hazards identification and risk management, safe operations, or conduct of operations • Address the reliability and integrity of the organization or assets
Content	Content considerations should include the following: • All PSM elements are adequately covered within the OEMS elements • All business processes are defined and appropriately addressed • All legal and regulatory requirements are covered
Structure	Structure considerations should include the following: • Cover with approvals • Introduction – purpose, application, and users • Roles and responsibilities • Requirements and expectations • Supporting processes, tools, and check sheets to achieve requirements and expectations
Ownership	Ownership considerations should include the following: • Assign ownership of elements based on organizational roles and responsibilities • Spread ownership and responsibilities to avoid overloading any one role (ensure expertise and leadership capabilities are strong) • Ensure commitment

4.2 Avoiding the One-Size-Fits-All Conundrum

Business Unit Differences

For integrated and diversified businesses, attempting to impose a one-size-fits-all OEMS process on all business units (BUs) is generally a fatal flaw. Unique business practices associated with BU and functional groups may require variants of the corporate management system. For most organizations in the energy industry, the following BUs are likely.

	BUs

Oil and gas	• Upstream
	  ○ On land
	  ○ Offshore
	• Midstream
	  ○ Terminals and pipelines
	  ○ Compressor houses
	• Downstream
	  ○ Refining
	  ○ Lubricants
	  ○ Upgrading
	• Marketing and distribution
	• Projects

Power and utilities	• Conventional power generation
	  ○ Coal-fired systems
	  ○ Gas-fired systems
	• Renewables
	  ○ Wind
	  ○ Solar
	  ○ Hydro
	  ○ Geothermal
	  ○ Battery Storage
	• Power transmission
	• Gas distribution
	• Water distribution
	• Projects

Functional organizations	• Human resources • Finance • Procurement • Legal and regulatory compliance

BUs and functions are often governed by different legal, regulatory, operating, maintenance, and reliability and integrity requirements and expectations. While the corporate management system strives to be all-inclusive, requirements and expectations may not necessarily apply to every BU within the industry.

Figure 4.1 provides a graphic representation of BU and functional organizations requirements within the corporate management system.

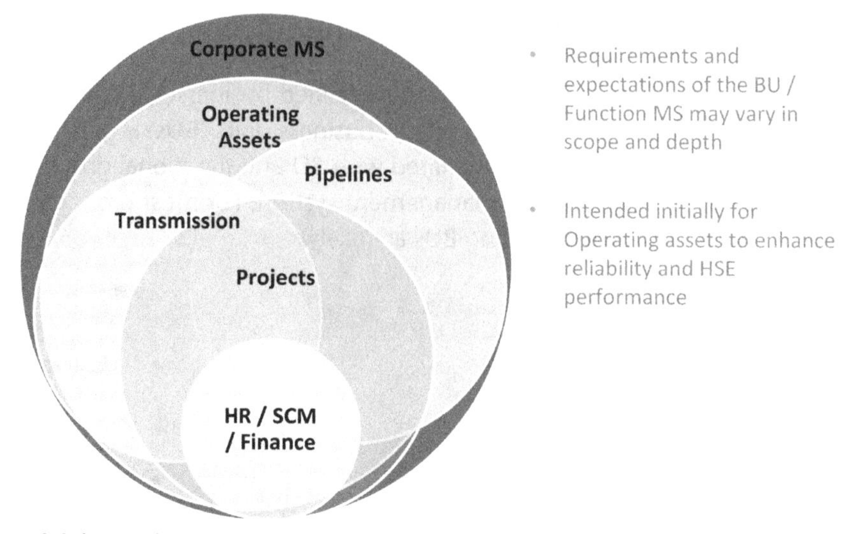

Source: © Safety Erudite Inc. (2019).
Figure 4.1: BU and functional organization differences in management systems.

4.3 Protecting Functional Organizations – HR, Finance, Procurement, etc.

Overview

Historically, OEMS was developed to protect organizations against major incidents associated with operating assets and high-risk facilities. The early origin of OEMS points to the Exxon Valdez incident of 1989 and Exxon's response with its 11 elements operations integrity management system (OIMS).

Since the introduction of OIMS, early adopters and subsequent followers failed to differentiate and evolve requirements and expectations to accommodate for BUs and functional organizations. Imposing the same unrealistic OEMS requirements and expectations across functional organizations results in undue stress and frustration within functional organizations in response to compliance audits. The outcomes of these unrealistic expectations often include the following:

- Blatant disregard for all requirements of the management system
- Selective compliance to cherry picked requirements and expectations
- Inadequate support to the dependent parts of the business
- Tension between personnel within the operating assets and functional organizations
- Other

Right Sizing for the Functional Organizations

While the requirements and expectations of some elements of the management systems – such as those related to leadership, training and competency, and emergency preparedness – are generally fully applicable to the functional organizations, others may require right sizing. Right sizing the requirements and expectations for functional organizations requires considerable involvement and engagement by the architects of OEMS with senior leaders (VP equivalent) of the functional organizations.

In the right-sizing process, functional leaders must

- Understand the requirements and expectations that are not applicable to the functional organizations and the reasons for them.

- Be able to recognize the demand implications of these requirements and expectations of the functional organization.
- Understand the challenges of misaligned or insufficient operations and maintenance needs from inadequate functional organization support.
- Be realistic in defining the requirements and expectations of the functional organization to meet the business needs of the organization

4.4 Defining the Required Element of the OEMS

Start with What You Have – Do not Reinvent the Wheel

When defining the requirements and expectations of each element of OEMS, the best place to start with is what you have within the organization. In acquiring multiple BUs, facilities and sites from different owners, multiple management systems documentation may be inherited. Inherited and legacy documentation relating to each element provides a basis for defining the requirements and expectations for each element. When defining the requirements and expectations, care must be taken to ensure the following:

- Documentation of the current status of the organization as a definition of the requirements and expectations of OEMS
- Documented requirements and expectations reflect the future state or are fully aligned with the OEMS vision for the organization
- Once completed and approved, all other source documents are retired and stored consistent with retention requirements defined in the documentation and information management standard

Legal and Regulatory Compliance Requirements

When defining the requirements and expectations of each element of the OEMS, the starting point for organizations is at a minimum shall include legal and regulatory requirements. For the energy, manufacturing, and most 24/7 process and continuous operations, there are various legal and regulatory requirements defined in regulations.

Learning from Industry Peers

Perhaps the most important source for defining requirements and expectations for each element is industry peers. The more mature the peer organization, the greater the accuracy of definition of requirements and expectations for most elements. In this context, maturity refers to a combination of the following two criteria: time since implementation of OEMS and business performance trends related to key Health, Safety and Environmental (HSE) and reliability indicators.

Leveraging Internal Subject Matter Expert Capabilities

Internal element-specific network leaders and subject matter experts (SMEs) are rich sources of knowledge for defining the requirements and expectations for respective elements. Rice-Bailey (2014) suggested, however, that knowledge from SMEs generated more value to users when restructured and presented by technical communication experts. SMEs are generally defined by experience, general knowledge, and competence associated with any particular area of the business. SMEs are known as the go-to persons for information on particular business or operating areas.

Leveraging Industry Associations, Forums, and Consultants

There are several industry institutions, associations, and forums that offer support expertise and documented information regarding each element of OEMS. Among them are as follows:

- Occupational Health and Safety (OH&S/OHS)
- National Transportation Safety Board (NTSB)
- International Association of Oil and Gas Producers (IOGP)
- American Fuel & Petrochemical Manufacturers (AFPM)
- OEMS Conferences and Workshops
- A growing abundance of OEMS consultants and consultancy services – care must be taken to verify the competency and capabilities of the consultancy services providers

4.5 Assigning Ownership of Elements

Ownership Based on Power

OEMS element owners may be assigned multiple elements based on discipline. For example, the HSE VP may be the owner of multiple elements as shown in Table 3.1. Assignment and ownership of OEMS elements generally lie with senior leader in the organization such as VP or department head. Success in implementation and sustainment of assigned OEMS elements is dependent upon both positional and personal power.

Source of Power	Implications
Positional	• Worker response primarily because of fear – work gets done because of the leader's position of dominance and power over the worker • Overuse of positional power leads to worker flight • Can get work done during periods of crisis
Personal	• Personal power is derived from moral authority derived from doing the right thing, ethical practice, and trust • Personal power inspires hearts and minds • Influence is limited based on positional power in the organization
Positional and Personal	• Provides strong ability to make things happen – work gets done because workers put extra efforts into getting things done • Followers are confident that positive change will occur and inject trust in the leaders/follower relationship

Creating Trust

Berens (2007) suggested "Trust makes the corporate world go round" (p. 1). Berens (2007) also advised a workplace replete with trust creates strong working relationships and a work environment where workers are focused on their jobs. There are fewer dysfunctional relationships to consume their time leading to higher and more sustained work performance. Trust is earned over an extensive period of demonstrated behaviors of doing the right thing. The ABCD of trust is a simple reminder to leaders on how trust is created in the workplace. As discussed earlier, success in assigning ownership of OEMS element and in its implementation is dependent on the leader's ability to create and sustain trust in the workforce. Figure 4.2 provides an overview of the ABCD model for creating and sustaining trust in the workforce.

ABLE -Demonstrate Competence:
- Produce results
- Make things happen – lead and manage change
- Know the organization: set people up for success

BELIEVABLE - Act with Integrity ... be Credible:
- Be honest in dealing with people: be fair, equitable, consistent, respectful
- Values-driven behaviors: generates confidence in followers, workers feel they can rely on their leader

CONNECTED - Demonstrate genuine care and empathy for people:
- Understand and act on worker needs: listen, share information, be a real person
- When leaders share a little bit about themselves, it makes them approachable

DEPENDABLE - follow through on commitments:
- Say what you will do and do what you say you will
- Be responsive to the needs of others
- Being organized, reassures followers

Source: © Safety Erudite Inc. (2019).
Figure 4.2: ABCD model for creating and sustaining trust in the workplace.

Desirable Attributes of Element Owners

When assigning OEMS element owners, care must be taken to ensure leaders possess the following attributes:

- Strong commitment to sharing the OEMS vision
- Trustworthiness
- Ability to make decisive decisions and promote continuous improvements
- Transformational leadership behaviors such as
  - The ability to inspire hearts and minds of followers
  - Create and steward strong teams
  - Lead and manage change
  - Excellent coaching and mentoring capabilities
  - Strong technical competence
  - Moral authority derived from doing the right thing, the right way all the time

4.6 Minimizing the Numbers of Elements Required

Overview

The fewer elements in the management system, the easier it is to steward and manage the implementation and progress. Key to success in any management system is as follows:

1. Effective management and control
2. Full coverage of all areas of the business activities

Fewer elements that completely cover all processes of the organization and correctly assigned ownership amplifies the effectiveness of OEMS. Careful attention should therefore be applied to these considerations. Figure 4.3 compares the number of elements of two different companies and how they are packaged.

OEMS Element – Company 1		Company 1 Elements - Aligned	OEMS Element – Company 2	
1	Leadership & Accountability	1. 6, 13, 16, 18	1	Leadership and Organizational Effectiveness
2	Risk Management	2, 15	2	Hazard Identification and Risk Management
3	Legal Requirements & Commitments	3	3	Working Safely
4	Goals, Targets, Planning	4	4	Environmental Stewardship
5	Management of Change	3	5	Legal and Regulatory Compliance
6	Structure, Responsibility & Authority	6, 14,	6	Asset Reliability and Integrity
7	Learning & Comp..	9	7	Conduct of Operations
8	Asset Life Cycle	5	8	Management of Change
9	Ops& Maintenance Control	7	9	Training, Competency and Human Performance
10	Contractor Management	10	10	Procurement - Contractor and Supplier Management
11	Data, Documents, Information Management	12	11	Emergency Preparedness, Security and Business Continuity
12	Emergency Management	11	12	Document and Information Management
13	Communications & Stakeholder Relations			
14	Quality Assurance			
15	Incident Management			
16	Audit & Assessment			
17	Corrective Actions			
18	Management Review			

Source: © Safety Erudite Inc. (2019).

Figure 4.3: OEMS elements equivalent to Company 1 and Company 2.

References

Berens, M. (2007). Trust and betrayal in the workplace: Building effective relationships in your organization. *Journal of Organizational Change Management, 20*(3), 463–465. Retrieved January 05, 2018 from emerald insight Database. doi:10.1108/jocm.2007.20.3.463.2.

Brien, F. O. and Meadows, M. (2000) Corporate visioning: A survey of UK practice. *The Journal of the Operational Research Society; Basingstoke, 51*(1), 36–44. Retrieved January 02, 2018 from ProQuest Database.

Llewellyn, E. P. (2005). Winning loyalty with a vision and a corporate soul. *The Health Care Manager, 24*(4), 374–378. Retrieved January 02, 2018 from EBSCOHost Database.

Rice-Bailey, T., (2014). Identity, value, and power: A qualitative study of the complexity of the working relationship between technical communicators and subject matter experts (order no. 3638903). Available from ProQuest *Dissertations & Theses Global. (1620743213).* Retrieved January 04, 2018 from ProQuest Database (inline image).

Safety Erudite Inc. (2019). The Integrated Process Management System Provider. Retrieved June 2019 from www.safetyerudite.com

U.S. Chemical Safety and Hazard Investigation Board (2007). Investigation report. Report no. 2005-04-i-TX. Refinery Explosion and Fire. Retrieved January 02, 2018 from www.csb.gov/assets/1/19/CSBFinalReportBP.pdf.

5 UNDERSTANDING DOCUMENTATION AND LEARNING TO WRITE FOR USERS

Overview

Documentation is among the most important elements of an OEMS. Properly written documentation is essential for defining requirements and expectations, processes, manuals, and many other communication requirements within the organization. Key tenets of an OEMS include the need for being organized, structured, disciplined, and *everyone doing the right thing the right way every time.* To achieve these tenets, adequate documentation and document management are essential.

In this chapter, the authors discuss the processes involved in developing, maintaining, and making accessible high-quality documents that are used within the organization. In many organizations, this will require a combination of standardized documents and consistent documents depending on the level of diversification of the organization. The goal is a transition away from checking the box that documentation exists toward disciplined and consistent use of all documents.

Doing the Right Thing the Right Way Every Time

Doing the right thing the right way every time will naturally vary from one individual to another. Therefore, it is necessary to help users with regard to the following:

- What is the right thing?
- What is the right way of doing things?
- What is required for getting it right every time?

Figure 5.1 demonstrates the concept of *doing the right thing the right way every time* as it relates to standards and procedures. Care must be taken to ensure that there are no errors of interpretation in documentation, so all users can *do the right thing the right way every time.*

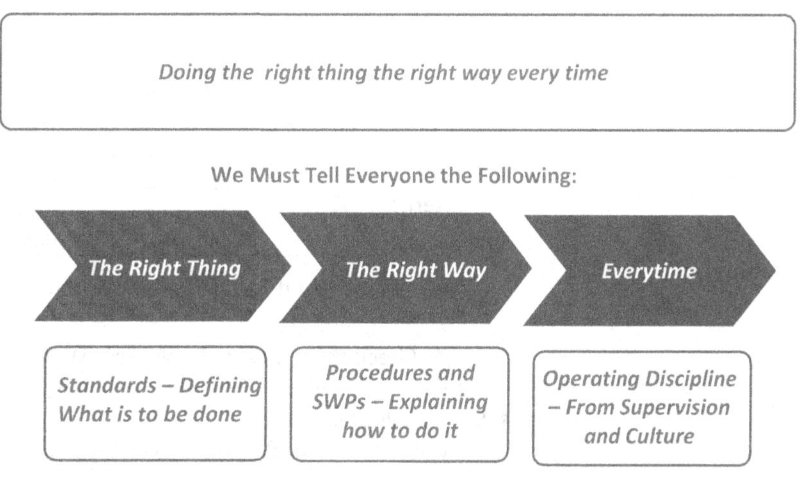

Source: © Safety Erudite Inc. (2019).

Figure 5.1: Doing the right thing the right way every time.

Standardized and Consistent versus Independence

Where OEMS is concerned, strong leadership is required to determine where standardization and/or flexibility are feasible. An overview of the differences associated with standardized/consistent processes versus doing the same thing differently at each facility or site of the business is shown in the following table:

Power Source	Implications
Standardized and consistent	• An overall low-cost producing organization • One set of documents to update/manage • Standardized for nondiversified businesses and consistent where diversification exists • Tendency toward a unified organizational culture
Different ways of doing the same thing (independence)	• Tendency toward a high-cost facility and a low-cost facility • Multiple sets of documents to manage and update annually • Unplanned internal competition that may be unhealthy as it relates to resources allocation and the halo effect • Tendency toward multiple cultures within the organization

5.1 Document Hierarchy and Document Types in Organizations

Hierarchy of Documents in Organizations

The hierarchy of documentation in organizations is important to all workers in communicating compliance or conformance actions as applicable. OEMS requirements and expectations are generally communicated in all levels of documentation in the organization. Figure 5.2 provides an overview of the typical document hierarchy found in organizations and the expected responses from all workers.

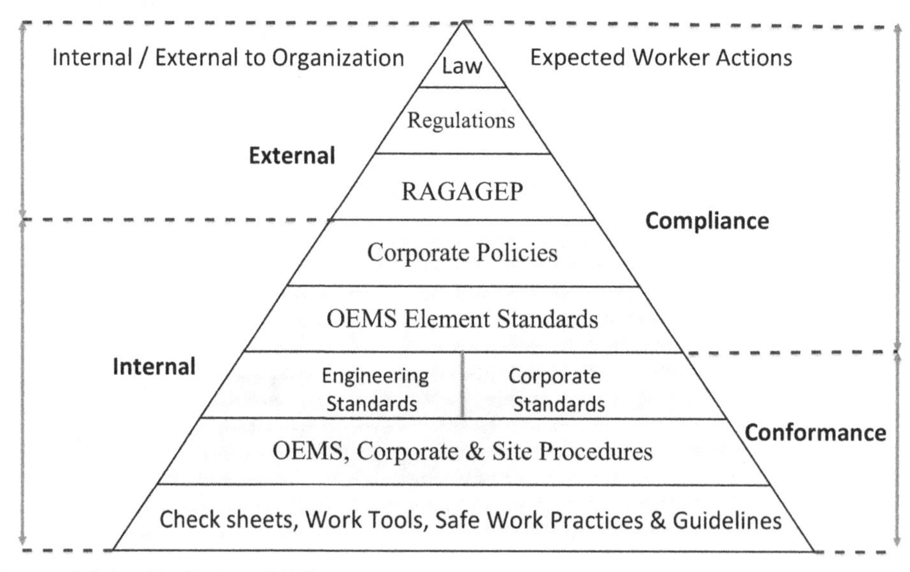

Source: © Safety Erudite Inc. (2019).

Figure 5.2: Overview of documentation hierarchy applicable to organizations.

Types of Documents in Businesses

All workers in the organization must be able to clearly differentiate the various types of documents in the organization and the intent of the document. More importantly, they must understand their expected actions relating to each document type. Table 5.1 provides an overview of typical document types found in organizations and the intent of these documents.

Table 5.1: Document Types and Intent

Internal Document Type	Description and Intent
Regulations	• External law that is generally quite extensive • Complete and full compliance is required
Policy	• Generally a one-pager to communicate the organization's position on a topic • Company law (e.g., harassment, EH&S, D&A, cellphone use while driving)
Standards	• Establishes the company's requirements for any particular discipline • Defines and communicates **WHAT** is required and expected
Procedures	• Provides a step-by-step process for achieving the requirements and expectations of the standard • Provides the **HOW** for meeting the **WHATs** in the standard
Engineering standards	• A requirement similar to a regulation or law that is provided for use as an industry standard, e.g., **American Society of Mechanical Engineers** (ASME) / American Petroleum Institute (API), etc. • Generally fully compliance or a documented variance required
Guidance documents	• Provided to help users better achieve the requirements of a law, regulation, policy or standard
Safe work practices (SWPs)	• Not to be confused with procedures, SWPs are available to assist workers in ensuring key activities/requirements associated with specific tasks are not missed and are addressed accurately • Cannot be used as a substitute for a procedure
Check sheets	• Provided to ensure company requirements or activities are not missed during the execution of work
Templates	• Provided for users to generate a consistent set of information required to steward or sustain a process

Source: © Safety Erudite Inc. (2019).

5.2 Documentation and Communications

Why Documentation and Writing

Documentation and writing provide the following opportunities and benefits:

- One of the most effective forms of communication
- Create a history or record of specific and general communication messages
- Provide a single source of truth
- Document expectations, requirements, and processes for doing something
- Replace your voice and directions when you are not there
- Allow the organization to meet a vast audience over any geographical expanse in a very short time

From an OEMS perspective, therefore, organizations should seek to provide users with consistent, quality documentation that is easy to use and apply.

Failures in Written Communication

Common failures in documented and written communication include the following:

- Written documents may sometimes be ambiguous and unclear when communicating instructions or guidance
- Written at a level inconsistent with the target audience
- Leaders may underestimate the value of face-to-face interaction when relying upon documentation to communicate critical information
- Many documents may lead to confusion from misunderstanding – difficult to follow, understand, interpret, etc.
- Over time circumstances may change, and documents may fail to evolve at the same rate
- Inconsistencies within the same document (e.g., policy versus standard versus procedure)
- Other reasons (unable to access, ignored, etc.)

These failures may eventually lead to frustrated users who eventually end up *doing their own thing.*

What Happens to the Frustrated User

When users are presented with documentation and writing that leads to frustration, the following are potential outcomes:

- The user proceeds to do what he/she thinks is right (use judgment) based on the knowledge and information accessible to them.
- In the absence of complete, accurate or up-to-date information, the user may often take a chance or wing it.
- The user falls back on the previous documentation and direction.
- Often a user may take no action or do nothing when a document appears to be inaccurate or insufficient.
- Most users may fail to seek help or additional information from a leader or source since it may make the user look bad or inadequate.

- Users lose respect for
  - The individual generating the document and communication
  - The cause (OEMS, PSM, etc.)

Perhaps, the most destructive outcome of these actions is an inconsistent organization culture reflected in different ways of doing the same thing and perhaps a growing number of incidents in the organization.

Attributes of Good Documentation and Writing

Desirable attributes of effective writing and documentation include the following:

- Accurate, simple, and relevant
- Easy to understand, follow, and use
- Written at the appropriate level of detail for the user
- Provides answers to the questions posed by users
- Users can find answers to their questions quite easily
- Where applicable, keep people safe and eliminate liability

5.3 Writing Effective Documents

Writing for the Workplace

It is simply not good enough to have documentation only that results in checking the box for complying with regulatory and business requirements. Documents should be written for users and not for writers. In the workplace, the types of documents typically used will include policies, standards, and procedures. The goal of document writing in OEMS is to drive operations discipline in the consistent use and application of documents. When writing for the workplace, document writers should seek to create finished documents with the following attributes:

- Should provide users the ability to do the right thing (from standards) the right way (from procedures) every time (operations discipline)
- At least three grade levels lower than the lowest qualification hired by the company within the group for which the document is intended
- Should anticipate the user's questions and answer them
- Provide cognitive linking such that users are always cognitively linked throughout the document with the topic of focus
- Should never compromise safety because of difficulties in interpreting what is meant by the information provided in the document

Removes Searching and Promotes Navigating

From the user's perspective, documentation should be provided that are easy to use and promotes navigation (as opposed to searching) to required information or knowledge. To demonstrate ease of use, consider the information provided in Figure 5.3. With ~144 words, when a user is asked to identify the *caution* applied when working with the film in question, the average user will take ~40–50 s to search for this information. At the first encounter of *caution*, the user stops searching and assumes he/she has found the information required and will generally stop at that point without reading further to see if any further caution is recommended.

Important Information

Store unopened film below 24°C (75 °F). Do not Freeze film. Use above 13 °C (55 °F), place developing picture in warm pocket. Warranty: Polaroid will replace film defective in manufacture, packaging or labeling. Warranty: Does not apply to outdated film and excludes all consequential damages except in jurisdictions not allowing such exclusions or limitations. Caution: This film uses a small amount of caustic paste. If nay paste appears, avoid contact with skin, eyes and mouth and keep away from children. If you get some paste on your skin wipe off immediately and wash with water to avoid an alkali burn. If eye or mouth contact occurs, quickly wash the area with plenty of water and see a doctor. Do not cut or take apart the picture or battery. Do not burn the battery or allow metal to touch its terminals.

Search 144 Words

Source: CAT-I Training (2014).

Figure 5.3: Searching for answers.

Consider now Figure 5.4, where the same information is provided (~142 words). When asked the same question regarding *caution*, the user finds the answers within 1–2 seconds and all of the *cautions* required are identified.

Title: Safety And Handling Guidelines

Storing	Always store your film as follows: • Away from Children and pets • In a dry place between 2-24°C
Developing	Pictures develop automatically Environmental temperature should be between 13-24°C Note: Develop in a warm pocket on a cold day
Caution	Please use caution because: • Corrosive: The film chemicals contain caustic • Electrical current: The film pack contains a battery
First Aid	First aid tips if a person comes in contact with developing chemicals in the film: **Chemical on Skin**: 1. Wipe off 2. Rinse with water **Chemical in Eye or Mouth**: 1. Wipe off 2. Rinse with water 3. Call a doctor
Warranty Policy	The warranty policy for this film is: **We Replace the Film if:** 1. Defective manufacture 2. Defective packaging 3. Incorrect labeling **We do not Replace the Film if:** 1. Outdated 2. Subsequently damaged
Disposal	Dispose of this film pack: • Like batteries • Do not incinerate

Search 142 Words

Source: CAT-I Training (2013).

Figure 5.4: Navigating to answers.

Outcomes of Searching – User Perspectives

From the user perspectives, the outcomes of searching are as follows:

Outcomes	Implications
Success	Finding what you are looking for
Dissatisfaction	Not finding what you are looking for Finding only part of what you were looking for
Distractions and loss of focus	Forgetting what you are looking for Becoming distracted by other items discovered while searching
Frustration	Giving up the search after some time
Errors	Thinking you found it when it wasn't what you were looking for

Introduction to Usability Mapping

In its simplest form, usability mapping is the process of mapping the user's questions to answers. CAT-I (2014) advised "Usability Mapping is a behavior centered industrial documentation strategy. This strategy is patterned around human perception, selection and response patterns …" (p. 3). CAT-I (2014) suggests that usability-mapped documents are designed to reduce risks arising from understanding documents and "foreign language interference" (p. 3). Figure 5.5 provides a list of user questions in what is called the margin titles with the answers to the questions in the promise to users section.

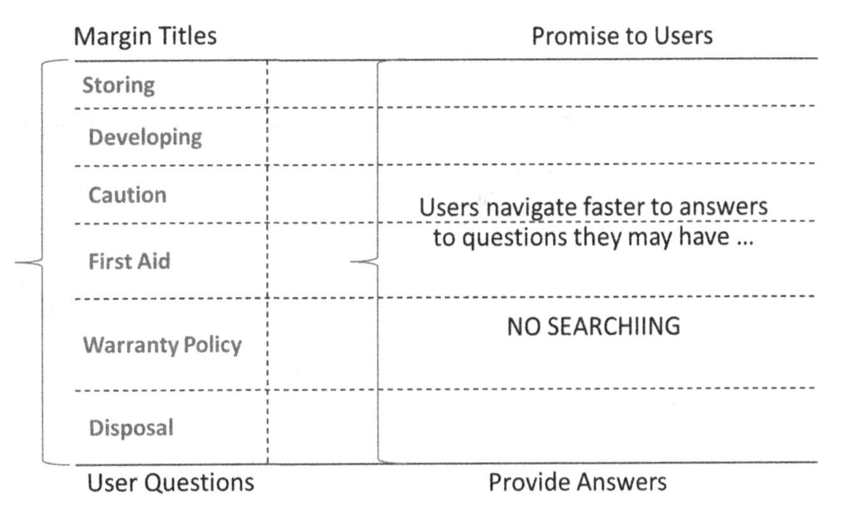

Source: CAT-I Training (2014).

Figure 5.5: Mapping user questions to answers.

Writing for Users

When writing for users, CAT-I (2014) provide the following guidelines to enhance usability:

- Write for three grade levels lower than the lowest qualification hired by the company to which the document applies – writing generally for a Grade 8 user or with a reading grade level (RGL) of ~8.0.
- Writers may reduce RGL by the following methods:
    - Converting sentences from passive voice to active voice
    - Reducing sentence length – rule of thumb: two lines
    - Reducing compound and complex sentences
    - Limiting the number of commas in a sentence to a maximum of four
    - Seeking to exclude semicolons
    - Using notations as follows:
        - Bullets with a range of 7 ± 2, i.e., 5–9
        - Roman numerals – <V
        - Alphabetical – a–h
        - Sequential – 1–7
        - Sequential – 1.0–5.0 (including sub 1.1–1.7)

Searching versus Navigating

In the writing process, the writer must be aware of the differences between searching and navigating to the user's questions. Figure 5.6 provides a summary of the differences between how users search versus navigate to answers in documents.

SEARCHING	NAVIGATING
• The User determine what to look for • The User searches for a credible hint within the areas where it could be or the User thinks it will be found • The User follows credible hints from one to the next until the search object is found	• Navigation is facilitated through cognitive links and hints that are supported with parallel constructs • These cognitive links and hints are strategically embedded redundancies in the writing to ensure the User can easily find the object
Searching is is regarded as guess-work in pursuit of credible *hints*	**Navigating involves following the *strategically imbedded cognitive links***
When a critical hint is missing, the chain of links is broken and the User switches back to searching	**Cognitive links provide confirmation to the User that he/she is on the right track**

Source: CAT-I Training (2014).

Figure 5.6: Searching versus navigating.

Cognitive Linking

Cognitive linking of the user's questions with answers is perhaps the most effective means of enabling users to navigate to the answers they seek in documents. Consistent with the guidance provided in Figure 5.6, user navigation is facilitated by strategically imbedded redundancies in the writing. Figure 5.7 provides an overview template of the cognitive linking of redundancies to help users navigate to answers where they are 100% confident that they have identified the right answer to their question or the information required.

Title: Safety and Handling Guidelines

Margin Titles	Promise to Users with Cognitive Links
Storing	Always ***store*** your film as follows:
Developing	Pictures ***develop*** automatically
Caution	Please use ***caution*** because:
First Aid	***First aid*** tips if a person comes in contact with developing chemicals in the film
Warranty Policy	The ***warranty policy*** for this film is:
Disposal	***Dispose*** of this film pack:
User Questions	Provide Answers

Source: CAT-I Training (2014).
Figure 5.7: Cognitive linking of questions to answers – Template.

Figure 5.8 provides a simplified sample showing the relationship between user questions (in margin titles) and answers to the user questions in which cognitive links are imbedded.

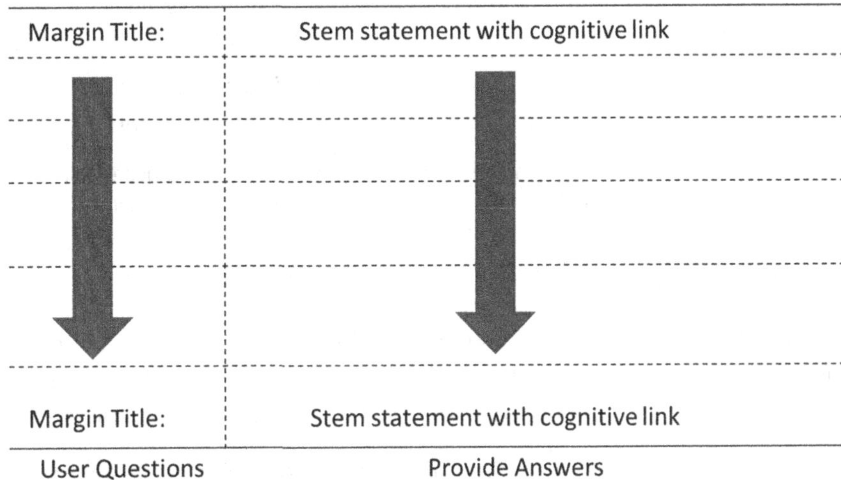

Source: CAT-I Training (2013).
Figure 5.8: Cognitive linking of questions to answers – Sample.

Types of Links in Documents

There are various types of cognitive linking in document writing. Writers must be able to provide users a clear path to what they need at all times during the use of documents. In view of this, the following types of cognitive linking may be applied to enable users to navigate to the answers to their questions or to the information sought. Table 5.2 introduces the various types of cognitive linking in documents to more effectively enable users.

Table 5.2: Cognitive Linking of Questions to Answers

Cognitive Link	Guidelines	Applies To
Question answer link	Writers should use the same words to provide a link between: • Topic titles • Margin titles • Stem statement • Content	All
Walking link	Writers should use the same word • To walk from the title to the process or procedure step • From a step of a process or procedure to "walk" to the next step • From the last step to "walk" to the closure statement • Often the linking word is near the end of the first step, and at the beginning of the next step	Process procedure
Navigation link	Writers should ensure that the same words (link) are used throughout your document for navigation as follows: • Table of contents • Chapter/sections • Topic titles • Margin titles • Stem statements	All
Title and closure link	• You should use the same words in the topic title and closure statements	Process Procedure

Source: CAT-I Training (2014).

Consistency in Documentation

Critical to writing documents is the need for consistency in referencing the various types of documents found in the workplace. It is not uncommon in almost every industry to encounter the following in documentation:

- Vast differences among worker definitions and interpretations of the various types of documents used in the workplace
- Within disciplines, writers may use the same term to define different things

Where these types of inconsistencies occur in documentation, immense confusion among users occurs and can often place unwanted liabilities unto the organization. Writers must therefore seek to ensure consistency when referencing the various types of documents and in the terminology used in documents. Figure 5.9 provides a graphic overview of the various types of documents and terms used in documents that, when used improperly by writers, may lead to user confusion and incidents.

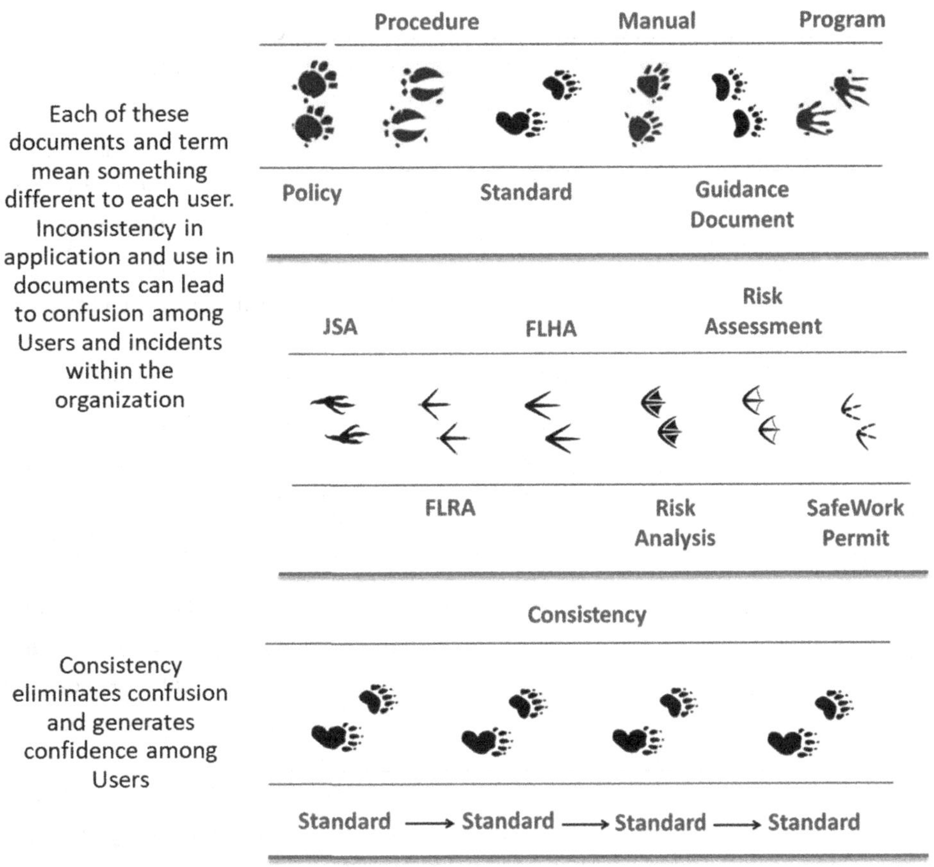

Procedure Manual Program

Each of these documents and term mean something different to each user. Inconsistency in application and use in documents can lead to confusion among Users and incidents within the organization

Policy Standard Guidance Document

JSA FLHA Risk Assessment

FLRA Risk Analysis SafeWork Permit

Consistency

Consistency eliminates confusion and generates confidence among Users

Standard → Standard → Standard → Standard

Source: Safety Erudite Inc. (2019).

Figure 5.9: Consistency in documents types and terms.

5.4 Creating Document Templates – Policy/Standards/Procedures

Why Document Templates?

Document templates help drive consistency in document writing. Templates provide document writers the benefit of knowing the types of information associated with the various types of documents. Key sections and content of the various types of documents are presented in the following table:

Document	Section and Content
Policy	• Governance and ownership section ○ Effective date ○ Policy owner ○ Policy approved by • Body and information section ○ Purpose of the policy ○ Scope ○ Applicable to ○ Consequence for violations

<table>
<tr><td>Standard</td><td>
<ul>
<li>Cover
<ul><li>Title</li><li>Owner</li><li>Approval</li><li>Originator and reviewers</li><li>Definitions</li></ul></li>
<li>Table of contents</li>
<li>Introductory section
<ul><li>Purpose</li><li>Objectives</li><li>Scope</li><li>Users of the standard</li></ul></li>
<li>Requirements of the standard
<ul><li>Follows the layout of the sub-elements defined in the OEMS framework</li><li>Roles and responsibilities</li><li>Health, Safety and Environment (HSE)</li><li>Other sections as required</li></ul></li>
<li>Supporting information</li>
<li>Tools and checklists</li>
<li>References</li>
<li>Revision history and records management</li>
<li>Appendices</li>
</ul></td></tr>
<tr><td>Procedures</td><td>
<ul>
<li>Cover
<ul><li>Title</li><li>Owner</li><li>Approval</li><li>Originator and reviewers</li><li>Definitions</li></ul></li>
<li>Table of contents</li>
<li>Introductory section
<ul><li>Purpose</li><li>Objectives</li><li>Scope</li><li>Users of the standard</li></ul></li>
<li>HSE requirements
<ul><li>Risk management</li><li>Personal Protective Equipment (PPE)</li></ul></li>
<li>Prerequisites</li>
<li>Procedure begins</li>
<li>Procedure review and revalidation</li>
<li>Revision history and records management</li>
<li>Appendices</li>
</ul></td></tr>
</table>

Figure 5.10 provides a typical template for developing a corporate or site-specific policy.

DRUGS AND ALCOHOL POLICY Policy No: 01 Total # of Pages:2

Company Logo

Governance and Ownership Section

Date Effective: Next Revision Date: Last Revision Date:

Policy Approved By:

Policy Owner / Manager:

Body and Information Section

Purpose, Application and User

Purpose of this Policy	The purpose of this policy is to ...
Requirements and Expectations	The requirements and expected behavior include ...
Scope	This policy applies to the following facilities / sites:
Applicable to	This policy applies to workers doing ...
Consequences for Violation	Consequences for violation of this policy include ... up to ...

Signature:

Source: © Safety Erudite Inc. (2019).

Figure 5.10: Policy template for consistency in documentation.

Figure 5.11 provides an overview of alignment of an OEMS standard with the OEMS framework element and sub-elements and a typical template for developing an OEMS standard.

OEMS Element 1- Leadership, Commitment & Accountability

Aims

OEMS Framework – Element 1

- Ensure all levels of management demonstrate leadership and commitment to operational integrity.
- Define and ensure appropriate accountability for operational integrity throughout the organization.

Sub-Elements – High Level Requirements and Expectations

1. OEMS is clearly articulated and communicated across the organization
2. Operational integrity management systems are established, communicated and supported at every level in the organization
3. Leaders visibly demonstrate commitment to OEMS
4. Responsibilities are clearly defined
5. The business environment is continually assessed for improvement opportunities
6. Internal learnings are incorporated into improvement processes
7. Expectations and requirements are documented and stewarded for compliance
8. Clear goals and specific objectives are established for each element of OEMS
9. OEMS performance is periodically accessed and communicated to all stakeholders

OEMS Element 1 Standard

- Introductory Section
  - Purpose
  - Objectives
  - Scope
  - Users of the Standard
- Requirements of the Standard
  - OEMS is clearly articulated and communicated across the organization
    - Detailed requirements and expectations
  - Expectations and requirements are documented and stewarded for compliance
    - Detailed requirements and expectations
- Supporting information
  - Tools and Checklists
- References
- Revision History and Records Management
- Appendices

Source: © Safety Erudite Inc. (2019).

Figure 5.11: OEMS element standard alignment and template.

Figure 5.12 provides an overview of the alignment of an OEMS procedure, standard, and framework for meeting the detailed requirements of the standard. There may be several procedures for meeting the requirements of an OEMS element. Table 5.3 provides differences between OEMS standards and procedures.

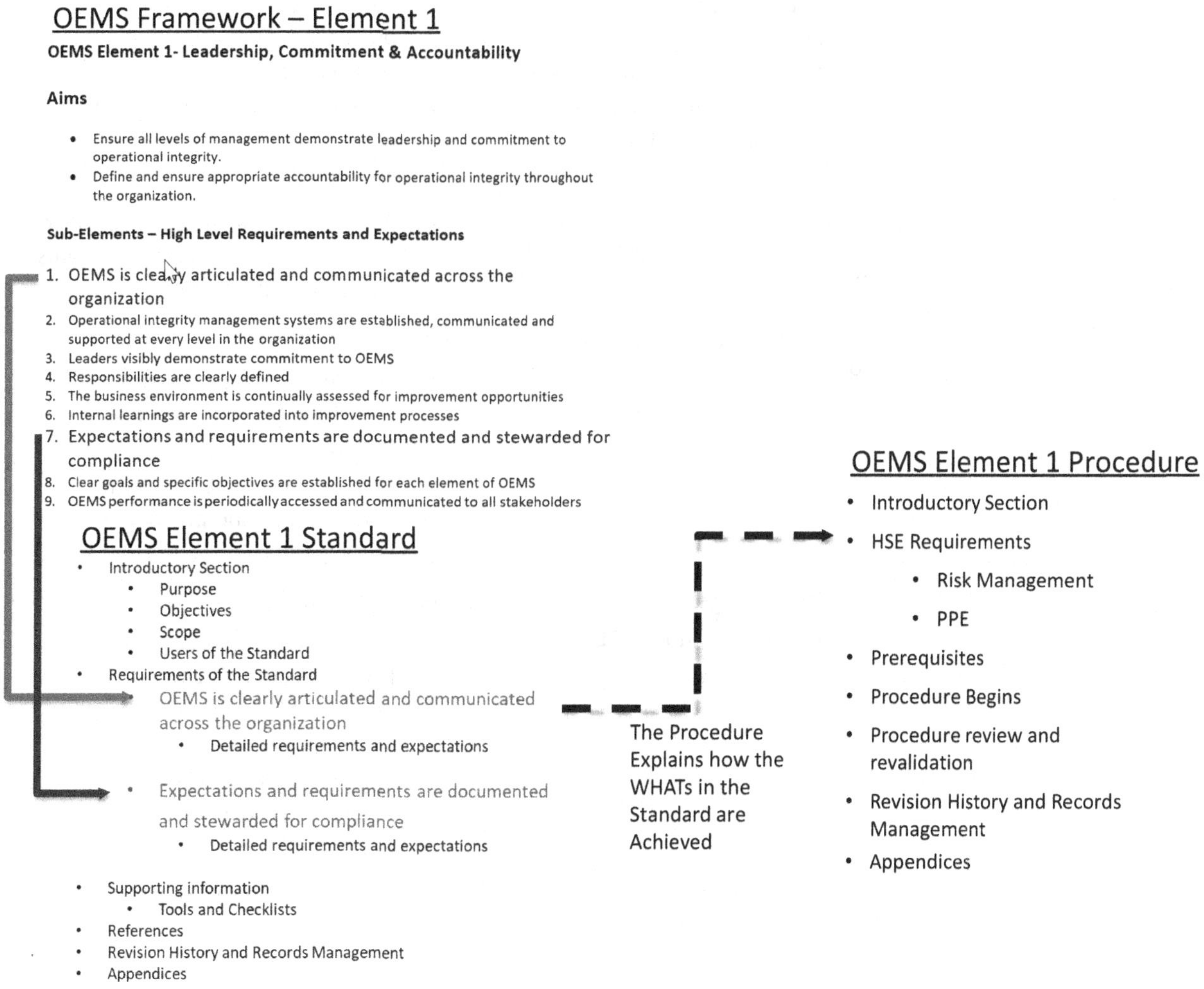

Source: © Safety Erudite Inc. (2019).
Figure 5.12: OEMS procedure alignment and template.

Table 5.3: Differences between OEMS Standards and Procedures

Standards	Procedures
Defines **WHAT** is required for each OEMS element	Defines **HOW** the **WHATs** identified in the standards are fulfilled
Established at the corporate level • Applied across all sites/ facilities • Right-sizing may be required for specific business areas and functional organizations	Established at the corporate level • Required where standardized practices are required, e.g., life-saving rules • May require site/facility-specific procedures for consistency as appropriate
Communicates the requirements and expectations of the organization to all sites/facilities	• Provides a step-by-step process for users to meet specific requirements and expectations
Ownership and change executed at the corporate level	Ownership and changes: • Executed at the corporate level where standardized practices are required • Executed at the site/facility level where supporting procedures are developed
Location and storage: • Corporate Intranet OEMS Home Page	Location and storage: • Corporate Intranet OEMS Home Page for corporate OEMS procedures • Site/facility-specific document management process where supporting procedures are developed

Source: © Safety Erudite Inc. (2019).

References

CAT-I (2014). Usability Mapping: A Learning Initiative for All. What Is It? – What Does It Cost? Retrieved January 13, 2018 from www.usabilitymapping.com/wp-content/uploads/2014/10/U-Map_BROCHURE_CAT-i_September_2014_v3.pdf.

Safety Erudite Inc. (2019). The Integrated Process Management System Provider. Retrieved June 2019 from www.safetyerudite.com

6 READYING THE ORGANIZATION FOR CHANGE – OEMS IMPLEMENTATION

Readying the Organization for OEMS

Readying the organization for OEMS is perhaps the most important stage in the implementation of OEMS. Careful attention to the change management process and adequate preparation is required to ensure success in this major undertaking.

OEMS represents a major change for most organizations. Even among many of the large and dominant organizations, the concept of OEMS is now taking roots within regarding the benefits and the volume of work associated with its development and implementation. Success therefore depends upon the use of a tried and tested change management model that may be appropriate for the undertaking.

Models for Leading the OEMS Change

There are various change management models adopted by organizations when readying the organization for change associated with OEMS implementation. Among these models that are considered for organizational and change associated with OEMS include the following:

- Kotter's Eight-Step Model of Organizational Change
- 7 Levers of Change
- 5-S Model of Change
- ADKAR (Awareness, Desire, Knowledge, Ability, Reinforcement) Model for Organizational Change

6.1 Kotter's Eight-Step Model of Organizational Change

Kotter's Model for Change

Kotter's model for change developed in 1996 is among the most widely used models for achieving organizational change. Wheeler and Holmes (2017) among several other users pointed successful application of this model in generating change. Kotter's model has been adapted for use in implementing OEMS and is demonstrated in Figure 6.1. The key steps in the process have been modified to reflect an OEMS perspective in creating and sustaining change.

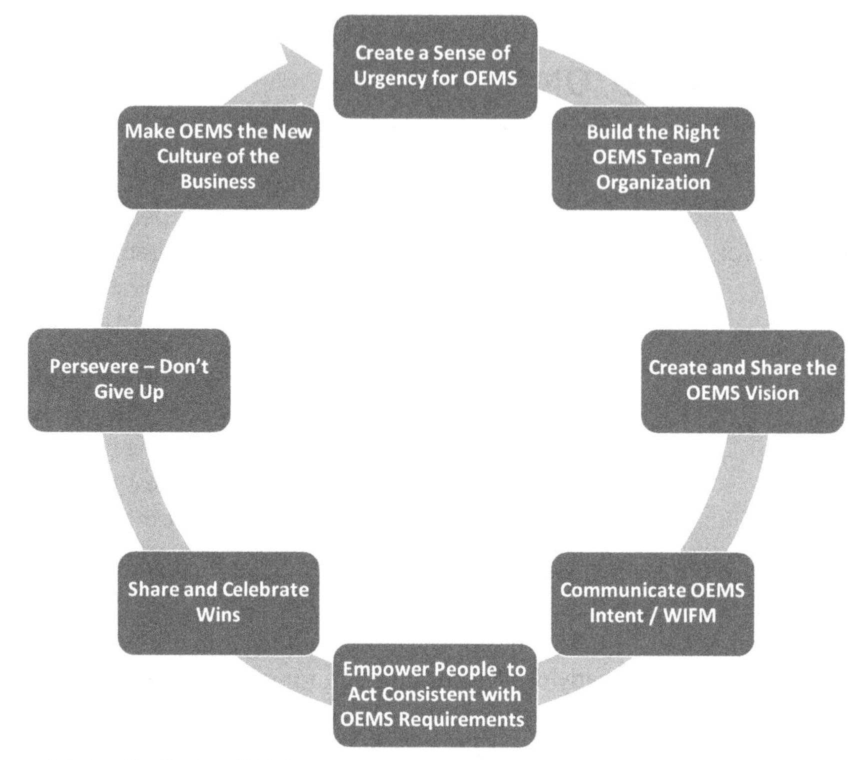

Source: © Safety Erudite Inc. (2019).
Figure 6.1: Kotter's Eight-Step Model of Organizational Change (adapted for OEMS).

Applying the Kotter's Model for Change

Calegari, Sibley, and Turner (2015) distilled the adoption of Kotter's model for building faculty engagement in accreditation identifying criteria for success at each stage of the process and how to achieve it. The authors of this book have adapted this work to provide a plausible model for generating organizational change in the implementation of OEMS.

The following table provides an overview of the goals, actions, and what should be done in each stage of the model as it applies to OEMS implementation.

Goal	How to Do It	What to Do
Creating a compelling sense of urgency and develop momentum for change	• Provide tangible, clear evidence of need for change • Reduce complacency and negativity • Appeal to specific interests and needs of recipients • Leverage advocates for change	• Compare with industry peers as a motive for change – Health, Safety and Environment (HSE), financial, and reliability performance • Highlight the *What's in it for me* (WIFM) for all levels of the organization – we all go home safely • Experts in various disciplines provide the value proposition for the proposed change – what will be better

Create a high-powered team to lead the OEMS charge – Form a team of powerful, influential, knowledgeable experts and develop a sense of teamwork and trust	• Provide the right incentives for team members • Develop the right organizational structure to ensure resources, visibility, and recognition • Encourage involvement from key extended participants • Reinforce involvement and collaboration – teamwork	• Select respected formal and informal leaders to lead the OEMS change/ implementation • Create trust in the team by ensuring team members are able, believable (speak from the heart and with compassion), connected, and can demonstrate the required behaviors
Create and share the OEMS vision – Create a vision and strategy that are aligned with the change	• Develop a concise OEMS vision and strategies that resonate with all levels of the organization • People should be motivated by the vision and want to be a part of the future • The vision can clearly guide the change effort	• Create a vision that emphasizes safety, reliability, and performance • Define strategies supportive of widely held values in the organization, e.g., creative thinking and questioning attitude • Strategies should generate confidence that the vision is achievable – behavior observation programs for improvement in HSE
Sell the OEMS vision and strategy – Enhance understanding and acceptance of the proposed changes so that new behaviors begin to reflect OEMS expectations and requirements	• Leaders who communicate OEMS are trustworthy and respected. They communicate with genuine care and empathy • First communication for all levels of the organization should be face-to-face and done by the CEO and the OEMS leadership team	• Senior and departmental leaders seize every opportunity to emphasize the value of OEMS – during team meetings and town hall meeting making OEMS a central theme • Highlight low-hanging fruit opportunities and success – areas where we can do better and areas where we derived success in the past, e.g., diligent application of Management of Change (MOC)

	• Ensure communication strategies are designed to minimize disruptions but effective enough to get the message across • Communicate frequently using multiple modes and avenues	• Dedicate resources toward accomplishing successful OEMS efforts to encourage participation and support and to demonstrate commitment • Make the communicated messages simple and easy for workers to grasp
Empower people to act consistent with OEMS requirements – Remove obstacles to action and reward behaviors that reinforce the change	• Ensure people know the right thing where OEMS is concerned – define them in standards and train users • Ensure people are trained on the requirements and expectations of the standards and supported with procedures • Create ownership of OEMS by expanding involvement and the range of participants	• Prepare an OEMS handbook for easy access and reference by users – particularly for those involved in implementation and those most impacted • Ensure the organizational structure includes an OEMS representative who can identify and eliminate barriers and obstacles, e.g., a senior OEMS coordinator who can act in this capacity
Share and celebrate wins – Support, communicate, reinforce efforts that are consistent with OEMS	• Define targets that trigger celebration of efforts toward OEMS and successes • Celebrate successes worthy of celebration. Celebration should be timely and relevant, and rewards should reflect the magnitude of the effort	• May include a letter from the CEO thanking everyone for their contributions (including those who did not participate ... this may get them on board as well)

	• Avoid celebrating everything, which may create the impression of a free-for-all	
Persevere and don't give up – Avoid second-guessing and recognize that OEMS is for the long haul. Focus on the big wins after the low-hanging fruits have been exhausted	• Compare OEMS requirements with actual situations – gap analysis • Prioritize gaps based on risk exposure, complexity, and efforts required • Have patience – give it time to work	• Continued communication (emails, internal web page, face-to-face meetings) on progress and implementation activities • Highlight the events that were avoided – high-risk gaps identified and addressed before an incident could have occurred
Make OEMS the new culture of the business – Make change stick whereby operations discipline behaviors become the norm	• Establish tangible improvements • Use orientations, training, other vehicles to instill/ reinforce change • Develop reward and recognition systems that institutionalize change • Align resource systems with change	• Imbed OEMS in orientation training for new hires to communicate OEMS expectations early • Recognize and reward operations discipline behaviors • Implement and sustain a president's award for operations excellence (PAOE) • Make OEMS a way of life • Assess the OEMS competency of leaders and tailor training programs for addressing deficiencies

6.2 7 Levers of Change Model

Overview

The 7 Levers of Change is a relatively undocumented model that has been very effective in leading and managing organizational change. Steps in this model include the following:

1. Leadership alignment and commitment
2. Stakeholder engagement
3. Organizational readiness and sustainability
4. Training and competency assurance
5. Communication
6. Setting up change networks and ambassadors for change
7. Enabling cultural change impact

When addressed, sequentially and systematically, this model is similar to Kotter's Eight-Step change management model earlier discussed. Figure 6.2 provides an overview of the model.

Source: © Safety Erudite Inc. (2018)
Figure 6.2: 7 Levers of Change.

Applying the 7 Levers of Change Model

Application of the 7 Levers of Change Management Model is summarized in the following table. This provides an overview of the goals, actions, and what should be done in each stage of the model as it applies to OEMS implementation.

Goal	How to Do It	What to Do
Leadership alignment – Develop leadership competence and capability at both the individual and group levels that will support the transition to operations discipline associated with OEMS	• Select effective and trusted leaders to lead and implement OEMS • Develop compelling, collective momentum for change • Develop and share the OEMS vision • Recognize that models of implementation may vary across different divisions and departments – adjust accordingly • Keep communication messages simple	• Involve the CEO in introducing OEMS across the organization and at all levels of the organization • Develop key and consistent messages and demonstrate behaviors that reflect commitment to the OEMS vision • Understand and leverage the corporate values and its alignment with OEMS • Resource the initiative fully

Stakeholder engagement – Strategies and interventions for various stakeholders at a group or individual level to support implementation of OEMS and operations discipline	• Prepare a change management heat-map to understand the magnitude and impact of the change • Conduct stakeholder impact assessment to determine approach for engagement • Develop strategies and interventions for various stakeholders at a group or individual level to move from the current state to required	• Develop and maintain a spreadsheet of all stakeholders that are impacted and assign risk mitigation strategies • Hold town hall meetings for input from stakeholders regarding how to most effectively implement OEMS • Address high-impact groups quickly – operations and maintenance organizations
Organizational readiness and sustainability – Conduct assessments of change effectiveness and behavioral change	• Identify key success factors that show the organization is ready to receive and sustain change associated with OEMS – People feel good about the proposed change and see a need for it • Complete the stakeholder impact assessment and address high-risk concerns • Perform surveys and focus group discussions	• Highlight incidents that were avoidable, near misses and lost production opportunities • Maintain and use an implementation readiness checklist to ensure all prerequisites are addressed • Perform HSE and culture surveys and address gaps identified in the readiness process
Learning and capability development – The development of skills, knowledge, and behaviors required for new ways of working. Aligned with centrally led learning and development methodology and programs	• Have a robust learning and development program to ensure people are competent for all assigned work • Assess the change impact of OEMS on jobs, conduct learning needs assessments	• Conduct AKSM competency analysis (A – Awareness; K– Knowledgeable; S – Skilled; M – Mastery) on OEMS and department leaders to ensure a competent organization

	• Determine OEMS training materials required and tools required to close skill gaps	• Develop and maintain a targeted training program to address deficiencies identified in core OEMS competency from the AKSM analysis • Check to assess OEMS-related training effectiveness
Communication – Communicate the need to do the right thing the right way every time	• Develop and implement awareness and targeted training to build initial awareness, understanding, and support • Communicate the right message from the right source to the right stakeholder using the right methodology – communicate frequently to get buy-in	• Develop and implement a communication strategy that is appropriate for the relevant stakeholder group • Communicate from the heart and with genuine care and empathy • Focus on sharing the OEMS vision higher up the organizational structure and the WIFM at lower levels of the organization • Prepare well and ensure consistent messages by all communicators – use a consistent set of speaking notes and key messages
Implement change networks	• Leverage existing network(s) to build momentum for change in operating areas and facilities • Build an integrated set of OEMS element networks with responsibilities for communicating and leading change associated with respective element	• Develop networks based on the guidance provided in Chapter 7

Culture – Align change activities and plan to promote and embed the operations discipline culture associated with OEMS	• Address gaps identified based on culture surveys – implement changes and reassess to determine the impact of change	• Be vigilant about walking the talk • Give it time to work – an OEMS cultural shift will take ~8–10 years unless there is a radical shift in culture

6.3 5-S Model of Change

Overview

The Five S (5-S) model for change was a five-step process developed in Japan is used in creating a culture that values organization and discipline. MCB UP Ltd (2003) advised this model is used primarily in creating a quality management culture and has been responsible for significant developments in the manufacturing sector; it is credited for business excellence in the manufacturing organizations. However, the principles are very transferable to OEMS implementation. Figure 6.3 provides an overview of the model.

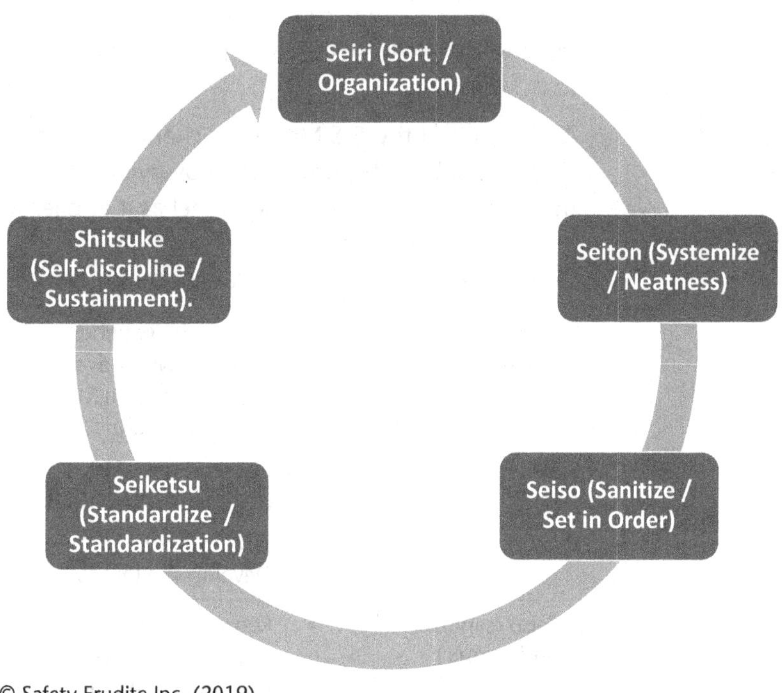

Source: © Safety Erudite Inc. (2019).
Figure 6.3: 5-S Model for Change.

5-S Model Explained

Nwabueze (2001) provided a practical application of the 5-S Model that is well explained in Figure 6.4.

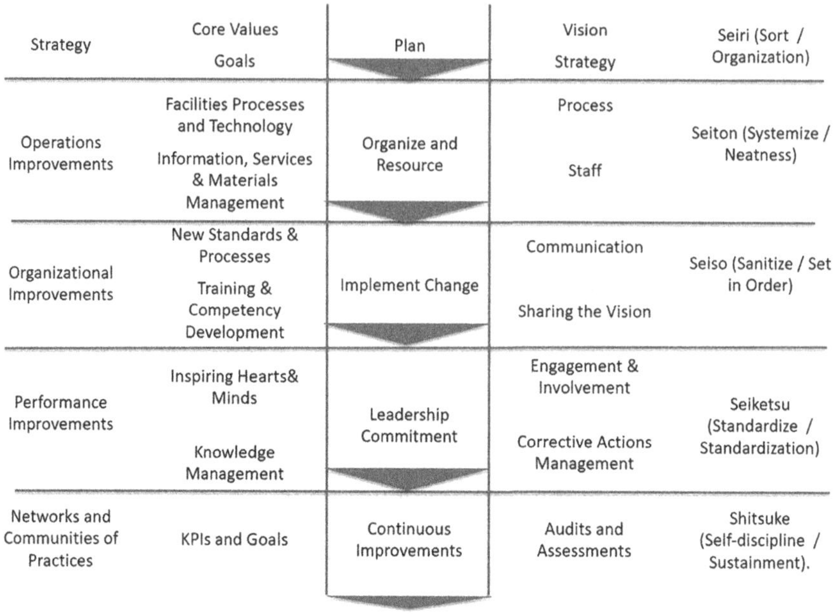

Strategy	Core Values Goals	Plan	Vision Strategy	Seiri (Sort / Organization)
Operations Improvements	Facilities Processes and Technology Information, Services & Materials Management	Organize and Resource	Process Staff	Seiton (Systemize / Neatness)
Organizational Improvements	New Standards & Processes Training & Competency Development	Implement Change	Communication Sharing the Vision	Seiso (Sanitize / Set in Order)
Performance Improvements	Inspiring Hearts& Minds Knowledge Management	Leadership Commitment	Engagement & Involvement Corrective Actions Management	Seiketsu (Standardize / Standardization)
Networks and Communities of Practices	KPIs and Goals	Continuous Improvements	Audits and Assessments	Shitsuke (Self-discipline / Sustainment).

Source: Safety Erudite Inc. (2019), Adapted from Prosci Research (2006).
Figure 6.4: 5-S Model for Change – a practical application.

Applying the 5-S Change Model

Application of the 5-S Model for change management is summarized in the following table. This provides an overview of the goals, actions, and what should be done in each stage of the model as it applies to OEMS implementation.

Goal	How to Do It	What to Do
Seiri (sort/ organization)	• Review the experiences of others who have implemented OEMS • Define elements, expectations, and requirements for the organization's OEMS • Develop the OEMS vision	• Leaders demonstrate the business vision, strategy, and implementation plan to the employees • Unleash the creativity of workers
Seiso (sanitize/ set in order)	• Reorganize the organization to meet the future state business needs – facilities, processes, and technology • Eliminate duplication and clutter – lighten the organization as applicable • Remove silos and improve integration among support functions • Develop standards and procedures for doing work	• Develop the organizational structure to best deliver on the OEMS vision • Activate the OEMS implementation organization • Develop OEMS networks to enable the OEMS process

Seiton (systemize/ neatness)	• Demonstrate business values and behaviors consistent with OEMS – operations discipline • Use standards and procedures when doing work • Learning good habits and steering people away from old habits	Leaders provide training to all employees regarding OEMS for understanding/ solving existing problems and leveraging opportunities Leaders consistently evaluate employee performance, skills, and behaviors to improve worker competency and motivate workers
Seiketsu (standardize/ standardization)	• Reinforcing the new way of working • Sharing knowledge and continuously evolving to find better ways of performing work	• OEMS capability development plans implemented for all workers • Networks fully activated and focused on resolving/eliminating OEMS gaps
Shitsuke (self-discipline/ sustainment)	• Continue breaking old habits and adhering to the new ways of working • Stick to rules, update standards and procedures, and provide training for all employees • Communicate frequently to ensure full buy-in and support for OEMS	• Share and reward success • Communicate progress in all levels of performance • Conduct audits and inspections to identify gaps and opportunities to continuously improve • Assess culture and take timely corrective actions

6.4 ADKAR Model

Overview

Prosci Research (2006), owners of the ADKAR change management model, provides organizations the secret to success in implementing and managing change in the ADKAR model. The ADKAR model provides a framework for managing the change dynamics of change at the individual. The ADKAR model represents the building blocks for individual change (Hornstein, 2015). A limitation of the ADKAR model is that it "fails to consider change to be a complex, systemic phenomenon that involves the interdependence of a multiplicity of variables and fails to highlight the important distinction between individual and organizational changes" pp. 295–296.

Figure 6.5 provides an overview of the model.

Pre-Contemplation Contemplation Preparation Action Maintenance

A D K A R

Awareness	Desire	Knowledge	Ability	Reinforcement
Communicate what is and is not working Define options available Define problems and solutions Focus attention on the most important reasons to change	Communicate benefits for adoption of change Identify risks involved Built momentum Address fears Define the WIFM	Effective training and coaching programs Job aides One on one coaching User groups and forums	Day to day involvement of supervisors Assess to subject matter experts Performance monitor Hands-on exercises	Celebrations and recognition Rewards Audits and performance monitoring systems Accountability systems Feedback from employees

Enablement Zone Engagement Zone

Source: Safety Erudite Inc (2019). Adapted From Prosci Research (2006).
Figure 6.5: The ADKAR change management model.

Applying the ADKAR Change Management Model

Application of the ADKAR change management model is summarized in the table that follows. This provides an overview of the goals, actions, and what should be done in each stage of the model as it applies to OEMS implementation. Mohan and Shubhasheesh (2017) provide an introduction of the ADKAR model as it applies to onboarding of new workers. Their process is adapted for the implementation of OEMS across organizations.

Goal	How to Do It	What to Do
Awareness – Create awareness with early communication among stakeholders about the business reasons for change (Prosci Inc., 2018)	• Maintain an effective OEMS introduction program for all workers impacted by the OEMS change • Ensure a credible and competent leader is available to build awareness across the organization	• Provide information about the organization's OEMS vision, mission, values, and culture • Communication and access to information are key for the awareness phase • Ensure timely provision of key information while managing information overload

Desire – Create interest and desire to be involved in the change – manage resistance (Prosci Inc., 2018)	• Improve or sustain motivation and reduce anxiety • Leaders should set the OEMS context right and explain the impact clearly to set expectations • This brings clarity on *What's in it for me?* (WIFM)	• Continue to inspire hearts and minds and reduce fear of the unknown associated with OEMS • Highlight opportunities and support – Identify and communicate access to information and knowledge • Highlight access to information and resources and people who can provide more information about OEMS
Knowledge – Enhance knowledge by training and coaching (Prosci Inc., 2018)	• Improve stakeholder knowledge by communicating, mentoring, training and coaching • Perform knowledge and competency gaps and take corrective actions • Ensure sufficient resources, adequate time, and timely feedback	• Share progress versus targets and goals • Verify processes and standards are implemented – Users are trained • Ensure full understanding on compliance required to organization's process and standards • Ensure full understanding and access to expectations and requirements associated with each OEMS element
Ability – Develop capability to implement change and generate performance (Prosci Inc., 2018)	• Workers are competent, apply knowledge, and demonstrate commitment to OEMS • OEMS is fully resourced • Inspire people – drive commitment and motivation	• Coach and mentor OEMS stakeholders to build OEMS abilities • Build commitment for the new way of working – a new culture is being developed

Reinforcement – Make change stick by measuring adoption, corrective actions management, and celebrating success (Prosci Inc., 2018)	• Hold people accountable and reinforce commitment • Manage and performance – recognize and reward behaviors and performance	• Link pay for OEMS performance • Communicate and celebrate performance • Introduce group/ team recognition and rewards programs

6.5 Why Change Fails – People and Organizational Change

Overview

Choi (2011) in a review of the literature pointed to the value of employee engagement in successful change management. He also identified the following attitudinal constructs that that must be addressed for successful change:

- Readiness for change
- Commitment to change
- Openness to change
- Cynicism about organizational change

Why People/ Organizational Change Fails

Figure 6.6 provides an overview of the essential requirements for ensuring people/organizational change associated with OEMS or any corporate initiative. When performed properly and detailed consistent with the earlier models provided the chances of success in people/organizational change is dramatically enhanced. When one or more components of the model shown are missing, people/organizational change is likely to fail.

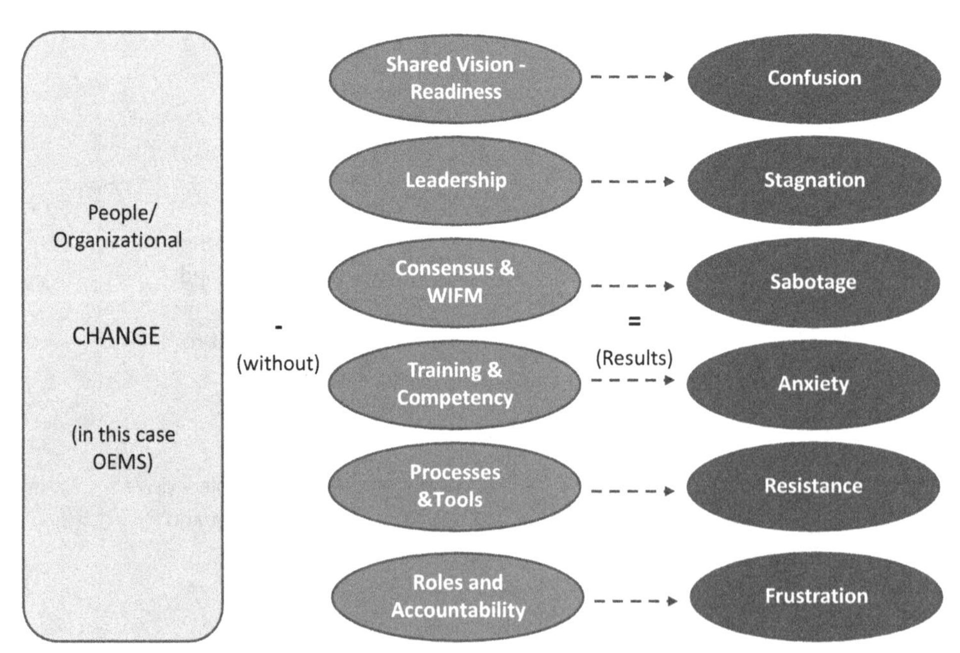

Source: Safety Erudite Inc. (2019).

Figure 6.6: Requirements for people/organizational change.

6.6 Assessment of the HSE and OEMS Culture of the Organization

Overview

As discussed earlier, most organizations have already begun the implementation of OEMS or its variants. When OEMS has been implemented and is not yet effective, unless the process is dramatically flawed, the best approach to improvement is to assess the effectiveness of various components of the management system and apply appropriate remedial actions. Figure 6.7 provides an overview of selected assessment criteria and suggested approaches to corrective actions. This list may vary based on organizational circumstances and leadership guidance regarding potential areas for assessments.

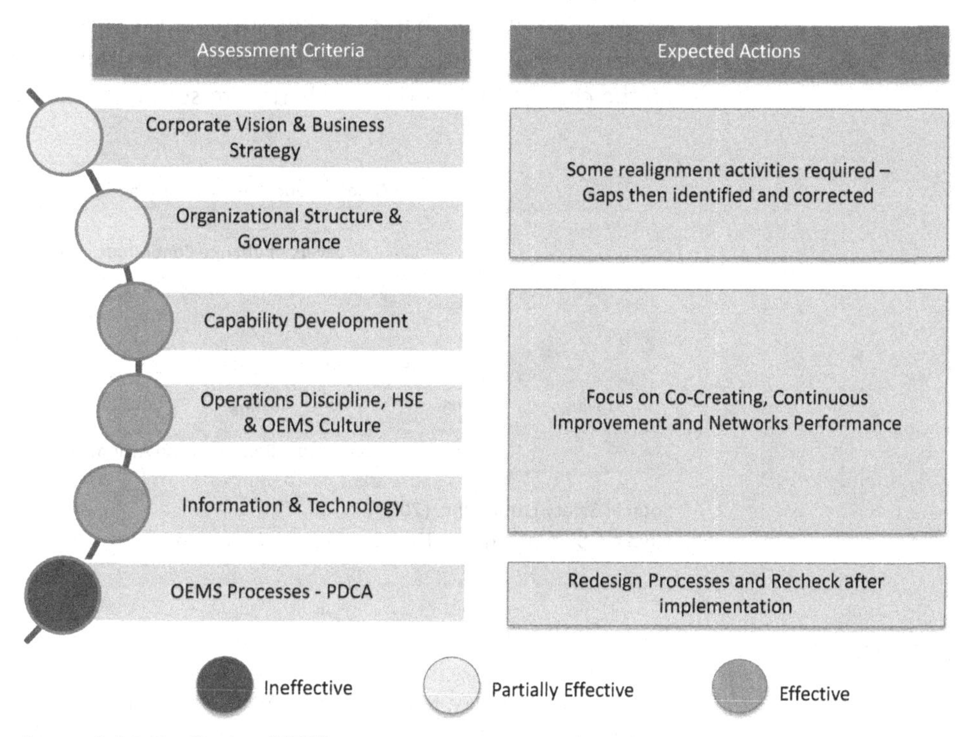

Source: Safety Erudite Inc. (2019).

Figure 6.7: Assessment criteria and suggested corrective actions for existing management system.

Evaluation Criteria and Effectiveness Assessment

In most instances, when assessing the effectiveness of OEMS implementation across an organization the strategy should consider the following guidelines:

- Evaluate effectiveness independently for each facility or asset (consider evaluating across departments when working with a nonintegrated organization)
- Consider findings as directional; seek trends upon which corrective actions should be based
- Corrective actions should range from process redesign to continuous improvements
- When impact is huge, consider using a change management model earlier discussed. Refer to Figure 6.8 for leadership actions based on the effectiveness status.

6.7 Selling the Recommended Implantation Model

Kotter's Eight-Step Model and the 7 Levers of Change Model

The authors are partial to Kotter's Eight-Step Model and the 7 Levers of Change Model for organizational change associated with OEMS. Using the concepts earlier discussed in this chapter, the authors are confident that when followed properly both these models have the capacity to deliver successful organizational change with OEMS.

Used properly and with the application of leadership actions defined in Figure 6.8, OEMS change can be sustainable.

Indeed, should any of the four change management models discussed be applied, the level of change and its sustainability may vary. Nevertheless, when implementing change, consider the use of one or multiple of these models based on the change focus – people change versus organizational change.

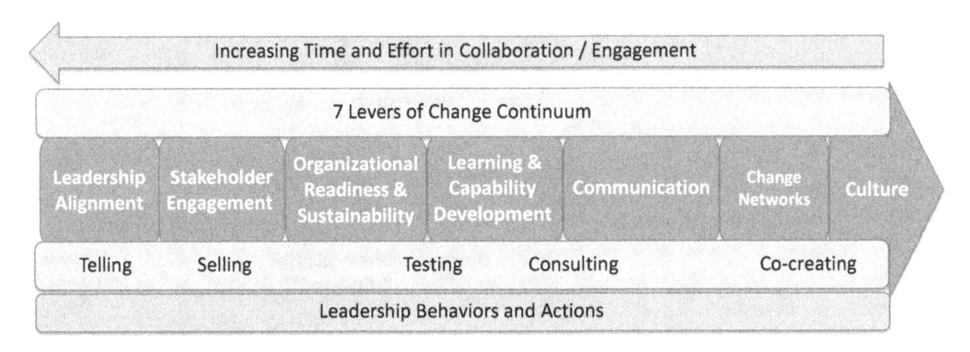

Source: Safety Erudite Inc. (2019).
Figure 6.8: Applying the change management strategy.

References

Calegari, M. F., Sibley, R. E., & Turner, M. E. (2015). A roadmap for using Kotter's organizational change model to build faculty engagement in accreditation. *Academy of Educational Leadership Journal, 19*(3), 31–43. Retrieved May 24, 2018 from ProQuest Database.

Choi, M. (2011). Employees' attitudes towards organizational change: A literature review. *Resource Management, Human Resource Management, 50*(4), 479–500. Retrieved May 26, 2018 from EBSCOHost Database.

Hornstein, H. A. (2015). The integration of project management and organizational change management is now a necessity. *International Journal of Project Management, 33*, 291–298. Retrieved May 25, 2018 from www.sciencedirect.com/science/article/abs/pii/S0263786314001331.

MCB UP Ltd (2003). MaST's 5-S journey to a smarter operation: Room for change in organizations. *Strategic Direction, 19*(1), 30–32. © MCB UP Limited 2003. Published by MCB UP Ltd. Retrieved May 24, 2018 from Emeraldinsight Database.

Mohan, K., & Shubhasheesh, B., (2017). Onboarding is a change. Applying change management model ADKAR to onboarding 2017. *Human Resources Management International Digest, 25*(7), 5–8. © Emerald Publishing Limited, ISSN 0967-0734. Retrieved May 24, 2018 from Emeraldinsight Database. doi 10.1108/HRMID-04-2017-0073.

Nwabueze, U., (2001). Chief executives hear themselves: Leadership requirements for 5-S TQM implementation in healthcare. *Managerial Auditing Journal, 16*(7). Retrieved May 24, 2018 from EBSCOHost Database.

Prosci Inc. (2018). PROSCI ADKAR MODEL, A Goal Oriented Change Management Model to Guide Individual and Organizational Change. Retrieved May 25, 2018 from www.prosci.com.

Prosci Research (2006). A Model for Change in Business, Government and Our Community. *PR Newswire; New York [New York]10*. Retrieved May 24, 2018 from EBSCOHost Database.

Safety Erudite Inc. (2019). The Integrated Process Management System Provider. Retrieved June 2019 from www.safetyerudite.com

Wheeler, T. R. & Holmes, K. L., (2017). Rapid transformation of two libraries using Kotter's eight steps of change. *Journal of the Medical Library Association, 105*(3), 276–281. Retrieved July 05, 2019 from EBSCOHost Database.

7 ESTABLISHING OEMS ELEMENT IMPLEMENTATION AND SUSTAINMENT NETWORKS

Overview

Lutchman, Evans, Sharma, and Maharaj (2013), drawing upon collective work experience of >100 years, documented the value of formal and informal networks and communities of practices in the workplace. They advised that networks contribute to continual improvement of business performance in the following ways:

1. "Learning from incidents that occur *within the organization* and developing practical and financially viable solutions to prevent repeat of the same or similar incident within the organization.
2. Learning from incidents that occur in organizations *within the same industry* and applying practical and financially viable solutions to prevent the same or similar incident from occurring within your organization.
3. Learning from incidents that occur *in other industries that can likely occur in your business or organization* and applying practical and financially viable solutions to prevent the same or similar incident from occurring within your organization.
4. From existing operations, *finding new ways and means to improve* the reliability, operability and performance of existing assets, technologies and processes." (pp. 1–2)

In this chapter, the authors explain the value of formal networks in the implementation and sustainment of OEMS.

How We Define OEMS Networks

According to Marshall and Simpson (2014), "Learning networks are typically depicted as collaborative settings capable of producing wise action" (p. 421). The authors suggest that OEMS and technical networks, in general, are teams of subject matter experts (SMEs) or like-minded individuals brought together to share knowledge, practices, and learnings so that we reduce risk, build competency, and accelerate continuous improvement in core business and focus areas. OEMS networks are defined as a core group of SMEs with responsibilities for implementing and sustaining the OEMS element across the organization. Where multiple sites exist, OEMS network members for each element may exist at each site. Typically, OEMS networks shall comprise the following:

- A network leader – responsible for coordinating the work of all assigned network members associated with OEMS for the organization or a particular site
- Network members – representation from each site, who will be the resident expert for the particular element and ultimately becomes the go-to person for that element
- A community of practice which serves as a resource pool of SMEs for the respective OEMS element

Networks Design

The typical design of an OEMS network is demonstrated in Figure 7.1. The OEMS comprises typically the following:

- A core group – A body of SMEs for providing expert knowledge relating to the network priorities
- A community of practices for supporting problems, leveraging opportunities, and helping with implementation at the field levels

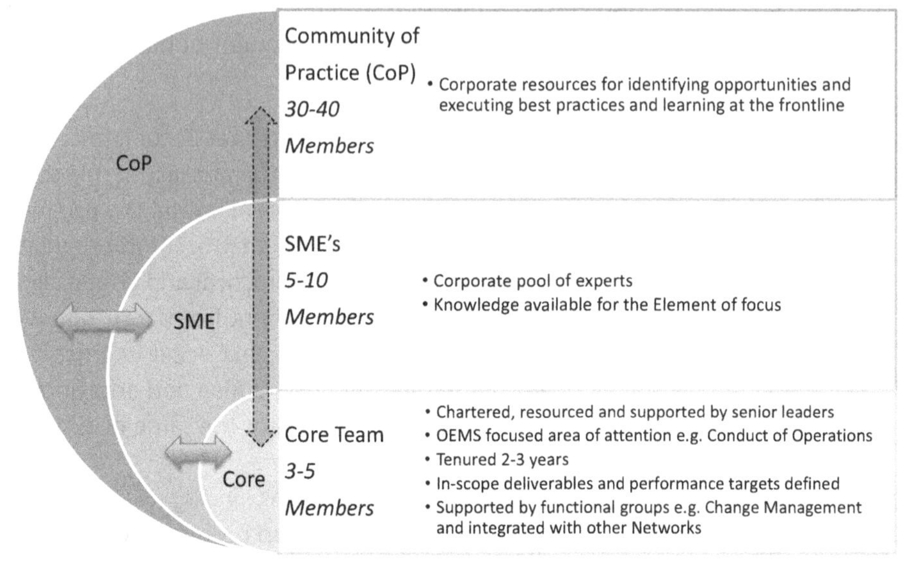

Source: Safety Erudite Inc. (2019).
Figure 7.1: Typical network structure.

7.1 Creating OEMS Ownership – Balanced Business Unit Representation

Building Ownership across Multiple Assets

A key attribute of OEMS is a drive for standardization and consistency across the organization in its business practices and the ways work is done. It goes without saying, therefore, where multiple assets or facilities are concerned (particularly where these assets are involved in the same types of work), each asset will want representation in determining how work is performed across all sites.

Building OEMS ownership is perhaps best achieved through involvement and representation for respective elements from each site. Ownership is best achieved when the following guidance is provided to network members and network leaders:

- Think big picture – focus on the organizational needs
- What works best for the organization versus what is best for my site
- Allocation of resources is based on risk exposures versus size of asset or profitability hierarchy
- Leveraging the work of all sites so that the best can be achieved for all
- Unbiased leadership – equitable and fair treatment to all network members by the network leader

Figure 7.2 provides an overview of the way OEMS networks work together to be successful.

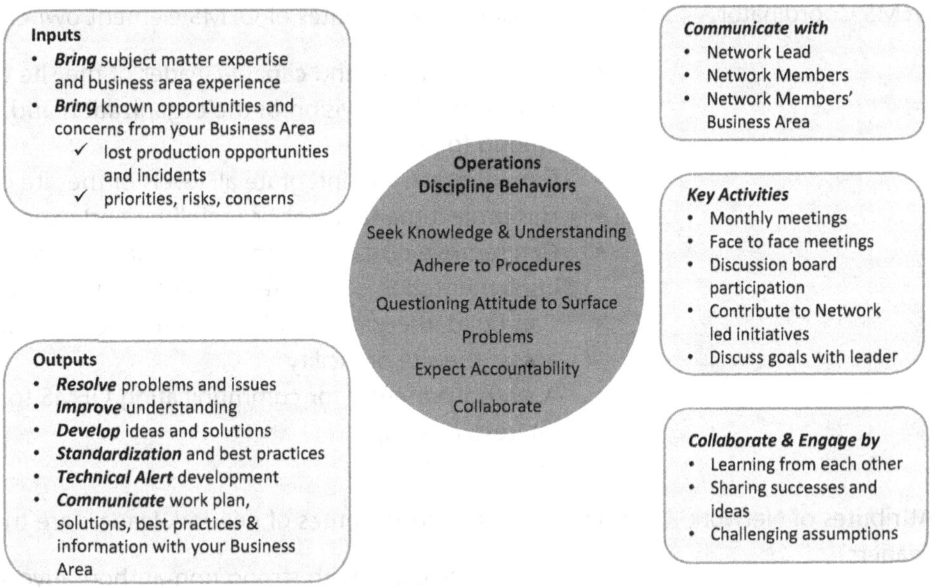

Inputs
- ***Bring*** subject matter expertise and business area experience
- ***Bring*** known opportunities and concerns from your Business Area
  - ✓ lost production opportunities and incidents
  - ✓ priorities, risks, concerns

Operations Discipline Behaviors

Seek Knowledge & Understanding

Adhere to Procedures

Questioning Attitude to Surface Problems

Expect Accountability

Collaborate

Communicate with
- Network Lead
- Network Members
- Network Members' Business Area

Key Activities
- Monthly meetings
- Face to face meetings
- Discussion board participation
- Contribute to Network led initiatives
- Discuss goals with leader

Outputs
- ***Resolve*** problems and issues
- ***Improve*** understanding
- ***Develop*** ideas and solutions
- ***Standardization*** and best practices
- ***Technical Alert*** development
- ***Communicate*** work plan, solutions, best practices & information with your Business Area

Collaborate & Engage by
- Learning from each other
- Sharing successes and ideas
- Challenging assumptions

Source: Safety Erudite Inc. (2019).

Figure 7.2: Overview of networks – ingredients for success.

7.2 Attributes of Core OEMS Network Members

How to Select Network Members

The success of OEMS implementation and sustainability depends on the quality and attributes of network members selected. When setting up OEMS networks, leaders should select carefully for the following attributes among others:

- Capacity for teamwork
- Shared vision – motivated and committed members
- Respected in the organization
- Can accept more workloads while handling core job responsibilities
- Balanced representation and SME capabilities and experience
- Can lead and implement work at respective sites
- Can leverage technology, and brings experience and knowledge to teams

Attributes of the various network members are provided in this section.

Attributes of OEMS Element Owner

Among the desirable attributes of OEMS element owner are the following:

- A senior, trusted, and capable leader of the organization
- Shares the OEMS vision of the organization and can continue to share it among followers
- Has access to, and control over, resources and decision-making
- Understands the strategic intent of the organization and can align element deliverables prioritization with strategic goals

OEMS Coordinator

Among the desirable attributes of OEMS element owner are the following:

- A senior, trusted, and capable leader of the site or facility
- Shares the OEMS vision of the organization and can continue to share it among followers
- Can interface and integrate all levels of the site organization
- Has project management capabilities and expertise
- Can remove obstacles within the organization to facilitate implementation of requirements of all OEMS elements
- Can build integration and collective involvement in implementing OEMS across the site or facility
- A strong presenter for communicating OEMS to large stakeholder and interested groups

---

Attributes of Network Leaders

Among the desirable attributes of network leaders are the following:

- Respected leader with strong non-authoritative leadership skills
- Can clearly articulate opportunities for standardization where applicable and drive consistency and standardization for the organization
- Decision-making capabilities
- Strong abilities to remove obstacles and interface with senior leaders of the organization
- Comfortable with technology and can create and lead strong virtual teams
- Can lead and manage change

Attributes of Network Members

Among the desirable attributes of network members are the following:

- Strong SME capabilities
- Company first ... not my plant
- Ability to work independently and take on new challenges and responsibilities with little supervision
- Strong people skills – site champion/coach/trainer
- Team player
- Possesses writing and presentation skills
- Comfortable with technology and virtual teams
- A DOER – willing to give more than required

7.3 Responsibilities of Core Participants and Structure of Networks Organization

Roles of core network organization core roles and responsibilities defined in the OEMS network organization are provided in Figure 7.3.

	Site 1	
	OEMS Coordinator • Coordinates Site implementation activities for all Elements of the Management System • Ensure full integration of all Elements of the OEMS for the Site • Prioritize implementation activities for all Elements of the MS • Removes obstacles and enables the implementation of all Elements for the Site • Provides site progress update on implementation of OEMS • Share implementation learning and successes across all Networks Members for the Site • Provide progress reports to the Corporate Organization on OEMS implementation and sustainment	
Element Networks	**Network Member**	**Network Leader**
Element 1 Network	• Be the site champion for Element 1 • Implement element requirements and expectations for the site • Capture and share learning within the Network • Provide SME knowledge and input within the Network • Work collaboratively within the network to find solutions that are best for the organization vs. what is best for my site	• Ensuring requirements and expectations are appropriate • Maintain Network focus on corporate goals associated with the Element • Generate continuous improvements of requirements and expectations • Shares learning among all sites via among Network Members and maintaining big picture focus • Select and steward high-value KPIs and drives performance improvements • Brings new knowledge to the Network from internal and external sources • Showcase Network activities, successes and tangible deliverables • Develop implementation plans for requirements and expectations for all sites
Element 11 Network	As Above	As Above
Element *n* Network	As Above	As Above

Source: Safety Erudite Inc. (2019).

Figure 7.3: Role of core OEMS elements network stakeholders.

Structure of the network organization across multiple sites and facilities – it is important that the right organizational structure is established for supporting the implementation and sustainment of OEMS. Figure 7.4 provides an overview of the recommended OEMS network structure for long-term implementation and sustainability.

Corporate OEMS Leader – Responsible for Implementation, Stewardship and Sustainment of OEMS across the entire Organization

	Site 1	Site 2	Site 3	Site 4	Site 5	Site n		
	OEMS Coordinator	OEMS Coordinator	OEMS Coordinator	OEMS Coordinator	OEMS Coordinator	OEMS Coordinator		
Element Networks	Network Member	Network Member	Network Member	Network Member	Network Member	Network Member	**Element Network Leader (NL)**	**Element Owner (EO)**
Element 1 Network	NM	NM	NM	NM	NM	NM	NL	EO
Element 2 Network	NM	NM	NM	NM	NM	NM	NL	EO
Element 3 Network	NM	NM	NM	NM	NM	NM	NL	EO
Element 4 Network	NM	NM	NM	NM	NM	NM	NL	EO
Element 5 Network	NM	NM	NM	NM	NM	NM	NL	EO
Element 6 Network	NM	NM	NM	NM	NM	NM	NL	EO
Element 7 Network	NM	NM	NM	NM	NM	NM	NL	EO
Element 8 Network	NM	NM	NM	NM	NM	NM	NL	EO
Element 9 Network	NM	NM	NM	NM	NM	NM	NL	EO
Element 10 Network	NM	NM	NM	NM	NM	NM	NL	EO
Element 11 Network	NM	NM	NM	NM	NM	NM	NL	EO
Element 12 Network	NM	NM	NM	NM	NM	NM	NL	EO

Element Owners – Responsible for Elemental / Discipline / Functional Progress and Reporting

Element Owners focus on the following:

- Strategically Improving Owned Element
- Developing annual work plans for Network
- Resourcing Network functions
- Participating in Network meetings and providing required Support
- Holding Network Members accountable for approved deliverables and Network performance
- Celebrating and sharing success

Notes:

- Element Network Members may serve on several OEMS Element Networks
- Element Network Leaders may lead multiple Element Networks particularly where there are synergistic opportunities
- OEMS Coordinators are Site Specific and cannot lead OEMS across Multiple Sites unless both sites are within the same location

Source: Safety Erudite Inc. (2019).
Figure 7.4: An effective OEMS network organization.

Network communication effective networks create a variety of communication methods for information and knowledge exchange such as peer-to-peer collaboration, expert input, face-to-face meetings, and events conferences (Leithwood and Ndifor 2016). Lutchman et al. (2013) advised on cost constraints in the business environment that drives the inclusion of virtual meeting places and virtual teams in the collaboration processes of networks. This is particularly important for global corporations.

According to Chao-Hua (2017, pp. 22–23), "high close-ness and frequency in contacts contribute to the accomplishments of complex collaborative projects in the vertical and horizontal learning networks."

7.4 Activating OEMS Networks – Network Conference

Overview

The benefits of an effective OEMS networks activation process cannot be understated. OEMS is perhaps the single most important initiative that will be undertaken in transforming the organization to operations excellence. Emphasis must, therefore, be placed on effective action of OEMS networks that are charged with the implementation and sustainment of OEMS across the organization.

There are various methods for activating networks, and among these are the following:

- Activation by element owner on a discrete element-by-element basis
- Activation by email from senior leader (corporate OEMS leader or president announcing the activation)
- Activation via an OEMS network conference

The effectiveness of each method varies. However, the most effective method appears to be via a network conference.

Pre-activation Preparation

Prior to activating the networks, preparation work shall ensure the following:

- Networks are chartered with defined deliverables
- A virtual network collaboration workspace is available to support the network work and activities
- Resources, support, guidance, and stewardship are in place
- Integration of existing management system (OEMS/Process Safety Management (PSM)/Health, Safety and Environment (HSE)) and technical networks to ensure full OEMS alignment

Why a Network Conference

A network conference is regarded as the most effective activation method because of the following reasons:

- When activated by the President/CEO of the organization, the importance of this work becomes paramount to all network members and involved parties.
- All members hear the same OEMS message regarding their OEMS role and responsibilities at the same time so that errors of chain communication are minimized.
- For diverse organizations with multiple assets, members are likely to have never met each other before. A conference provides opportunities for members to meet each other for the first time.
- All elements of OEMS are integrated and network members are required to work together. A conference provides opportunities for members of different elements to integrate and interact with each other.
- A network conference may be costlier initially; however, cost savings from performance improvements from effectively functioning networks more than offsets the cost of network conferences.

Key Participants of Network Conference

Key participants at the network conference should include the following:

Who	Why
President and CEO	To ensure the importance of OEMS is cemented in the hearts and minds of participants
Keynote speaker – respected leader in society	Typically, someone who is reputable and admired in society or in the country – to inspire hearts and minds
External OEMS/network experts	To share experiences and provide knowledge on networks and OEMS
Element owners	To ensure the following: • Roles and responsibilities are properly understood. • Clarification is sought as required. • Network peers and other network members are met. • The team-building process is initiated.
Corporate OEMS leaders	
All members of the OEMS network organization	
Leaders of technical networks (and members if cost permits)	

7.5 Criteria for Selecting and Activating New Networks

Networks Objectives

The principal objectives of networks are as follows:

- To define best practices
- To benchmark key performance indicators (KPIs)
- To create and share knowledge and learnings
- To identify and address opportunities for risk reduction and business performance improvement

Network Selection Criteria

Once the value of networks is properly articulated and success is shared, there is a general tendency toward getting everything done through networks. The concept of small teams, work groups, and group meetings appear to disappear into networks. Too many formal and informal networks may overly burden the organization. A mechanism for determining whether a network is required is therefore essential.

The role of network screening and selection is best managed by the OEMS stewardship organization whose function must include prevention of the unnecessary formation of networks (Figure 7.5).

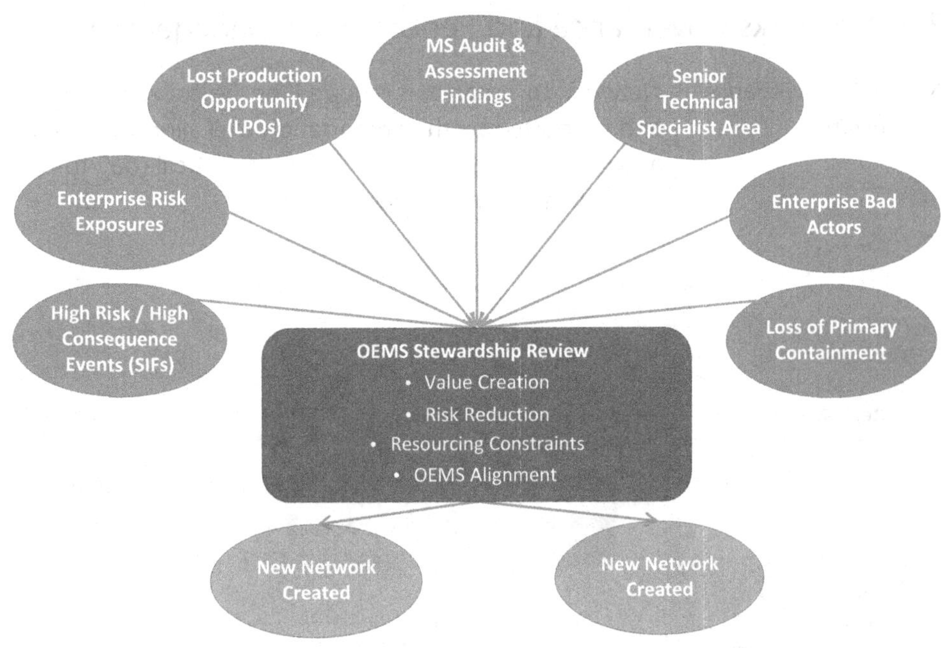

Source: Safety Erudite Inc. (2019).
Figure 7.5: Criteria for screening and selection of new networks.

OEMS and Technical Networks in the Oil and Gas Industry

Typical OEMS and technical networks in the oil and gas industry are provided in Figure 7.6. These networks can be developed and activated when required.

Disciplines			
MS (OEMS)	**Engineering**	**M&R**	**Operations**
PSM and HSE	Process Technology	Maintenance & Reliability	MS Networks
• **Elements of the MS** • Risk Management • Management of Change • Learning & Competence • Asset Life Cycle • Safe Operations • Contractor Management • Information Management • Incident Reporting & Investigations • • **Element *n***	• Hydro-processing • Fluid Catalytic Cracking (FCCU) • HF Alkylation • Energy Efficiency • Water Treatment • Process Simulation • Distillation • Delayed Coking • Process Automation Naphtha Reforming • Sulphur/Amine	• Maintenance Execution • Planning and Scheduling • Turnarounds • Reliability • Rotating Machinery • Process Automation Systems • Electrical • Inspection • Materials • Pressure Equipment and Piping	• Operating Procedures and Safe Work Practices • PSSR • Process Hazards Analysis • Operating Envelopes • Incident and Emergency Management • Operator Training and Competency

Source: Safety Erudite Inc. (2019).
Figure 7.6: Typical OEMS and technical networks in an oil and gas organization.

7.6 Networks Governance and Performance Management

Network Governance Organization

Network governance is required for the collective success of all networks created and supported by the organization. Full alignment between the strategic goals and vision of the organization to the tactical (day-to-day) work performed at the frontline is required (Figure 7.7).

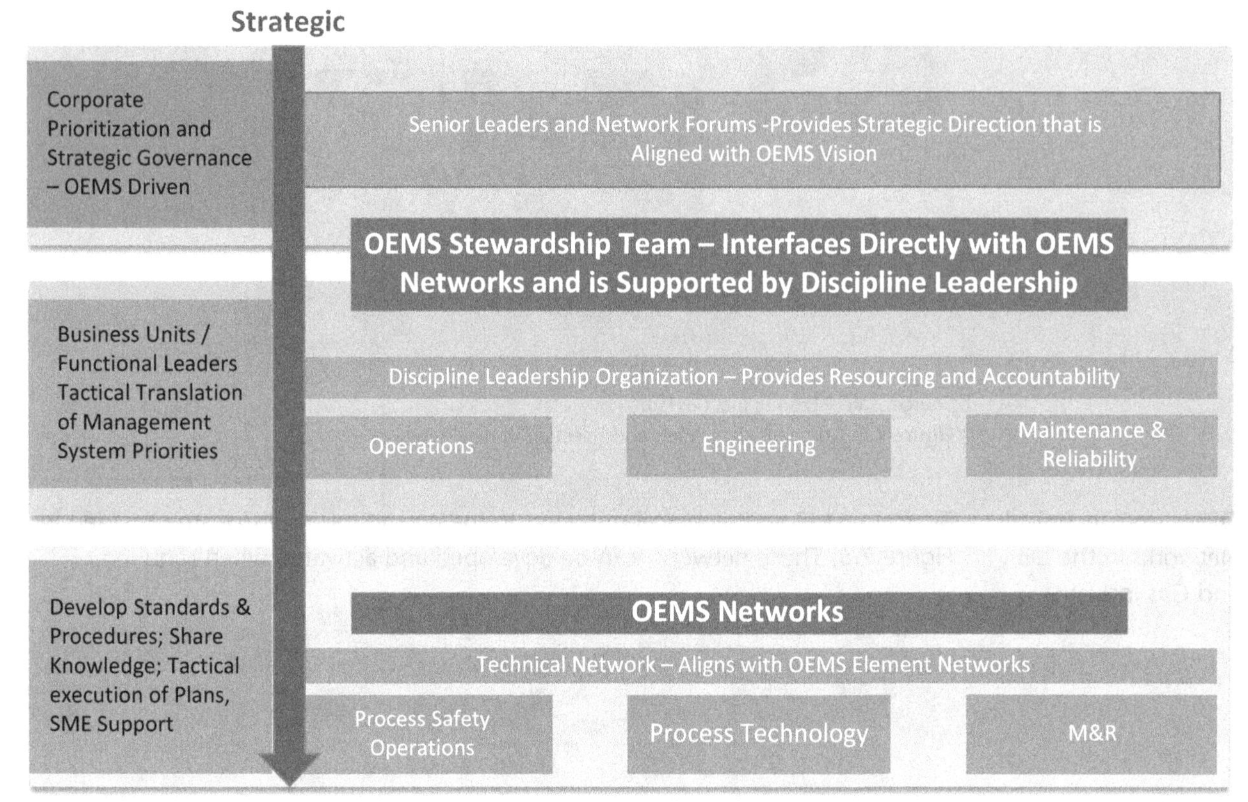

Source: Safety Erudite Inc. (2019).
Figure 7.7: Network governance structure.

Work Coordination

Coordinating the work of all networks is a complex process that requires attention and collaboration among respective stakeholders. Strategic goals must be delivered from tactical plans that are generated from iterative discussions among the stakeholder groups as shown in Figure 7.8.

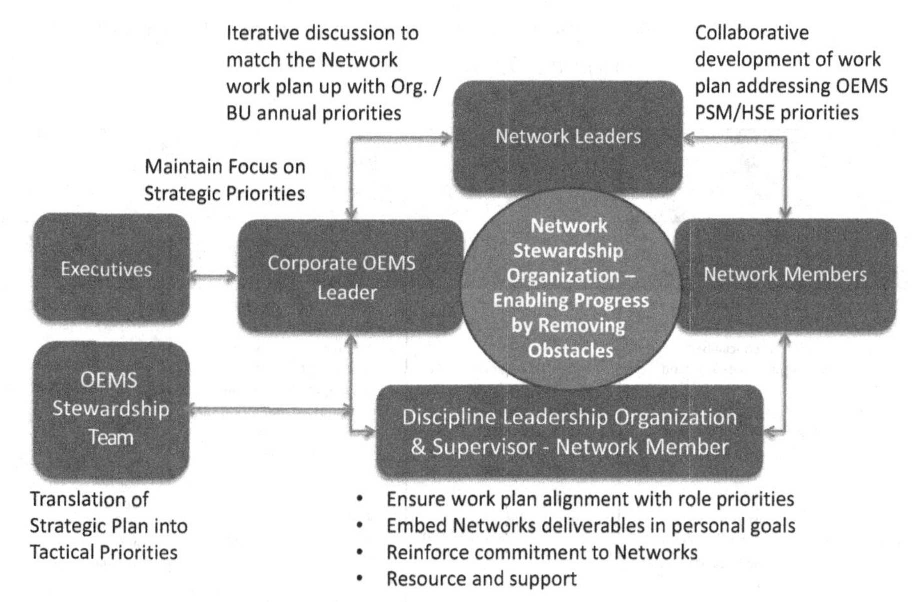

Source: Safety Erudite Inc. (2019).

Figure 7.8: OEMS and tactical network coordination.

Managing Networks

Effective management of OEMS and technical networks requires consideration of the following:

- A networks charter that defines scope, deliverables, and commitments
- Defined network success factors/performance indicators (KPIs)
- Attach networks to closely related departments (network owner department)
- Include network performance KPIs in owners' department performance monitoring systems
- Review network performance on a periodic interval
- Assess network performance and make timely changes to network management when necessary

Network Support Required

Effective network engagement helps to achieve the following:

- Support and achieve the shared (management system) vision
- Create buy-in and support adoption
- Provide avenues for stakeholders to surface relevant ideas and issues
- Help us get there faster and in a more sustainable manner

Chartering Networks

Chartering both OEMS and technical networks helps to maintain focus on prioritized network deliverables for the organization. "Priorities should be kept simple, concise and clear, and limited to a small number of achievable goals at a time" (Leithwood and Ndifor, 2016, p. 413) allowing correct decision-making and direction setting. A network charter also provides a framework for keeping all network deliverables moving forward in a systematic manner.

A sample OEMS network charter for management of change (MOC) is provided in Figure 7.9.

Purpose: To ensure that everyone involved in Management of Change (MoC, excluding MoC-People) understands and apply MOC effectively to deliver operational excellence. We leverage the knowledge and expertise that resides within the organization and industry.
Timeframe: Steering team to review network effectiveness monthly ***Commencing:*** (insert Activation Date Here)

Objectives	Measurable Goals	Key Activities	Tracking Measures (How)	Critical Success Factors
o Support/drive operational excellence in management of change (MoC). o Compare, contrast, reconcile best internal operating practices between sites across HFC o Identify areas of opportunity for continuous improvement of the MoC Standards o Support implementation of the standards across the network o Provide a forum to facilitate these objectives efficiently and effectively o Provide guidance to the network on the interpretation of standard o Build competencies required to perform MoC in the communities of practice (CoP)	o Effective usage of MoC processes across the organization o Increased conformance to MoC standards across the organization o Maintain and continuously improve Standards by regularly engaging the network. o Timely response to questions / requests from the business	o Monitor the current state of MoC across the organization o Define the MoC metrics for improving MoC performance o Drive the improvement of the standard, definitions and audit protocol documents o Seek out best practices from across organization and industry for organizational sharing o Participate in PSM Audits relating to MoC as needed. o Interface and collaborate with other networks	o Audit and self-assessment of performance o Improvements in performance on corporate and local MoC KPIs & metric o Reduction in turn around time for support / responses to queries from business and CoP o Demonstrated behaviors endorsing the use and application of the MoC Standard and its intent.	o Results oriented. o BU approval and support of members commitment Effective steering / facilitation o Excellent communication and visibility. Clear performance measures o Membership and performance of the members in their personal goals and scorecards o Best in class infrastructure to support efforts o Collaborating with other networks
In Scope	**Out of Scope**	**Deliverables**	**Members**	**Meetings**
o Assist in revision and continuous improvement of MoC Standards o Development and enhancement of training material (generic – not site specific) o Engagement with the network covering all business units/areas o Key stakeholder for MoC (IT) Tool development / sustainment o Inclusion of Management System requirements for Management of Change	o MoC-People. o Site Implementation of MoC or other site specific initiatives o Site selection of MoC personnel. o MoC IT tool implementation and sustainment	o An up-to-date MoC Standard, audit protocols, PSM definitions o Up-to-date training materials o Monthly stewardship summaries of activity and results	o Sponsor – o Team Lead o Team Members	o Notionally 12 meetings per year – 1 face-to-face annually – 11 virtual meetings per year (1-3 hours each) o Initial face-to-face meeting of all members - 2 days. o Monthly update to steering team

Source: Safety Erudite Inc. (2019).
Figure 7.9: Sample OEMS network charter – MoC.

Managing Network Performance

Managing the performance of networks is the responsibility of the network leader. Monitoring the performance of all networks is the responsibility of the network stewardship team and ultimately the accountability of the corporate OEMS leader. When managing network performance, the following guidance is provided:

Results	Expectations
Quantifiable	• Contribute to value creation at least ten times the cost of running the network • Profit improvement • Expense reduction or cost avoidance
Qualitative	• Use risk/consequence approach to define value for risks mitigated or avoided • Organizational competency, efficiency, alignment, standardization, and shared learnings

Performance Monitoring

Monthly monitoring of network performance is required to identify and eliminate obstacles to network performance. Figure 7.10 provides a simple template for documenting and reporting network progress. Obstacles are easily identified and resolved with the help of various stakeholder groups.

Network Monthly Update - Template

Network – MoC	**Date:**
Leader:	
Overall Status	o On target o Small manageable delays o Needs help from steering team
Sub Initiative Updates(s)	o Clearly identify the status of network activities and priorities generated from the opportunity matrix. o Provide updates on all sub-initiatives being worked by the network
Next Period Outlook	o Provide a listing of all major upcoming activities projected by the network

Support Requirements	
Network Stewardship Organization	o Identify support required from the steering team by the network
Issues and / or Challenges	o Provide a list of challenges faced by the network that may require senior leadership or business unit leadership support for addressing them

Network Successes
o List all achievements and successes delivered by the network

Network Participation	
Meeting attendance	
% of network absenteeism	
# of meetings cancelled	
# of network vacancies	

Source: Safety Erudite Inc. (2018).

Figure 7.10: Monthly element network progress update report.

References

Chao-Hua, L. (2017). An exploration study of organizational learning networks: An identity approach. *International Journal of Organizational Innovation. Hobe Sound, 10*(1), 13–26. Retrieved June 01, 2018 from ProQuest Database.

Leithwood, K., & Ndifor A. V., (2016). Characteristics of effective leadership networks. *Journal of Educational Administration, 54*(4), 409–433. Retrieved June 01, 2018 from doi:10.1108/JEA-08-2015-0068.

Lutchman, C., Evans, D., Sharma, R., & Maharaj, R. (2013). *Process Safety Management: Leveraging Networks and Communities of Practice for Continuous Improvements*. CRC Press.

Marshall, N., & Simpson, B. (2014) Learning networks and the practice of wisdom. *Journal of Management Inquiry, 23*(4), 421–432. Retrieved June 01, 2018 from ProQuest Database.

Safety Erudite Inc. (2019). The Integrated Process Management System Provider. Retrieved June 2019 from www.safetyerudite.com

8 PRE-IMPLEMENTATION PREPARATION AND READINESS

Overview

Readying the organization for change is a critical component of the overall change in management process. Careful attention is required to ready the organization for this change. As shown in Figure 6.8, leadership alignment is the first step in readying the organization for OEMS and the required change. In this chapter, the authors provide a pathway for readying the organization for the required change by leveraging the 7 Levers of Change Model discussed in Chapter 6.

At this stage of the process, the following components of OEMS are now in place:

- Leadership and cultural alignment.
- Elements of the management system have been defined.
- Expectations and requirements for each element have been defined in standards – the cumulative integration into a single document now represents the OEMS Manual.
- Supporting standards, procedures, tools, and templates for each element have been identified and are either in place or being developed.
- Element networks have been identified, though not yet activated.
- Collaboration workspaces for networks are under construction or are largely developed.

Overview of the Readiness Components

Figure 8.1 provides an overview of the preliminary stages of OEMS readiness for launch.

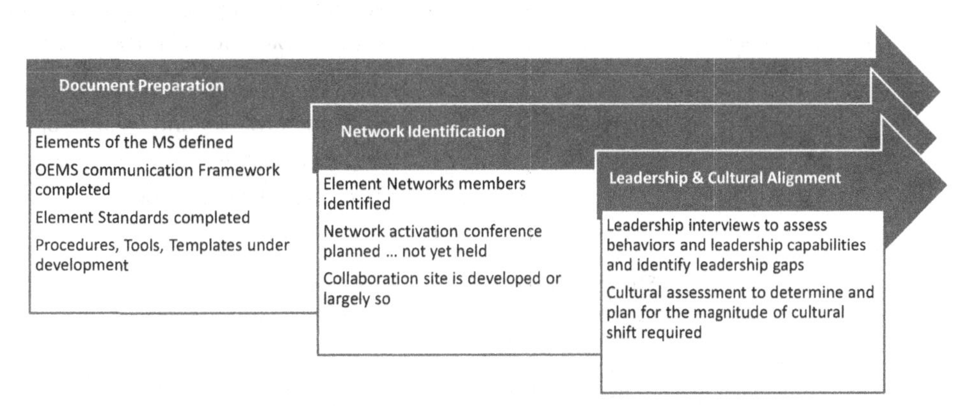

Source: Safety Erudite Inc. (2019).

Figure 8.1: Overview of the preliminary stages of OEMS readiness for launch.

8.1 Leadership and Cultural Alignment

Desired Leadership Behaviors

Getting leaders aligned with the vision and goals of OEMS is necessary for successful implementation and sustainment of OEMS. Leadership behaviors best suited for OEMS implementation and sustainment that are also characteristic of high-reliability organization are detailed as follows:

Behavior	What We Mean
Questioning attitude	• A questioning attitude helps spot what others may have missed • A questioning attitude forces you to maintain a sense of chronic unease in the quest for better ways of doing things • A questioning attitude promotes regular attention in searching for and identifying improved methods • Are vigilant that the information they receive may be truncated and or distorted • Constantly asking what else can be done to solve problems and remove barriers
Integrity and courage	• Integrity equates to reliability and is directly correlated to ethical behavior • Doing the right thing takes courage and enhances integrity • Those with integrity actively work to gain the perspective of others • Feedback is acted upon • Takes the necessary risks • One's word is kept • Those with integrity do not partake in or allow finger-pointing
Engagement	• Tremendous knowledge and value are derived from engagement, and networks have been developed to enhance the value of engagement • Involves asking for and offering feedback • Uses positive accountability to cultivate a culture of trust within the organization • Requires meeting regularly with staff to *actively listen*, and asking probing open-ended questions
Increasing knowledge	• Knowledge provides the basis to question and to develop out-of-the-ordinary and creative opportunities and solutions • We see knowledge beyond our own area to expand our ability to back up our coworkers • We are consummate learners
Structured approach	• Following the process to execute work and to generate continuous improvements
Disciplined	• We follow the process/procedures with consistency
We create trust	• We create trust in the way we work • We follow the ABCD of trust • We are: A. *Able*: Demonstrate competence in performing duties B. *Believable*: Act with integrity C. *Connected*: Demonstrate genuine care, empathy, and concern for workers D. *Dependable*: Demonstrate reliability in following through on promises

Understanding the Existing Culture

Understanding the existing culture of the organization helps in the successful implementation and sustainment of OEMS. A cultural assessment (consistent with the approach demonstrated in Figure 6.7) is done to understand the following:

- The status and maturity of the existing organizational culture
- Attributes of the organizational culture that are effective, somewhat effective, and ineffective
- Opportunities to correct weaknesses in the existing organizational culture that may impede the implementation and sustainment of OEMS

Leadership Alignment

Leadership alignment requires an understanding of the current state of leadership at all levels of the organization relative to the behaviors defined earlier in this section. Structured interviews are conducted with open-ended questions to identify gaps, if any, relative to required leadership behaviors defined above. Findings provide opportunities for the development of targeted training programs to close gaps identified before the implementation of OEMS.

8.2 Communicating and Sharing the OEMS Vision

Defining the OEMS Vision

Defining the OEMS vision is necessary for sharing it. Considerations when defining the vision are as follows:

- The OEMS vision must be simple and should resonate with employees at all levels of the organization.
- Generally stated in one sentence – two lines at most.
- Employees should easily relate to the vision and should want to be a member of the team making the journey toward the stated vision.
- There should be no ambiguity in the vision – it should be clearly articulated and achievable.
- The vision should enable all employees to be a part of the process in achieving the vision.

Communicating and Sharing the OEMS Vision

Communicating and sharing the vision is not a simple process. Careful attention of the following is required:

- Communication and delivery method
- Who delivers the message – a trusted senior leader of the organization is required
- In-person versus electronic or printed materials
- Approach to sharing the vision

Among the most effective approach to sharing the vision is that demonstrated in Figure 8.2.

Path of Development

A Look Back – 3-5 Years | Where We are Today | OEMS Vision – Where we want to be in 5-8 Years

What we looked like 5 yrs. ago

- Multiple sites ... different cultures
- People dependence
- Reactive and inconsistent
- No basis for resources allocation
- Good practices ... no documentation
- Repeated incidents
- Good efforts to improve processes that eventually went away
- Resistance to sharing *–not invented here*
- Duplicated efforts
- No setting of expectations
- Weak accountability

What we look like Today

- Multiple sites ... different cultures with smaller differences
- Less dependence on people – growing dependence on processes and systems
- Less reactive but still inconsistent
- Growing trends towards allocating resources based on risks exposures
- Good practices ... no documentation
- Fewer repeated incidents ... transitioning towards sharing and learning from each other
- Silos being broken down
- We are working together more closely

Our Vision 5-8 years from now

- Multiple sites all operate safely with common purpose, culture and pride
- People matter, but we have the structure to carry on when they leave
- Consistent and proactive
- Thoughtful risk management across the company
- Auditors audit us against our standards and procedures
- We learn from incidents when we have them and share with each other
- Processes are consistent where they need to be and customized where they need to be, but no matter what we sustain and improve them
- We demand sharing and collaboration
- We leverage our scarce resources – divide and conquer
- Mature Networks – the heavy lifting will be done, we will be sustaining and improving
- Everyone knows what is required of them to run safely and reliable

Source: Safety Erudite Inc. (2019).
Figure 8.2: Approach to sharing the OEMS vision.

Using Multiple Communication Channels

Every opportunity must be used to share and remind workers of the OEMS vision and how we get there. The most effective method for doing so is, of course, face-to-face discussions. However, his method is extremely expensive particularly for geographically distributed organizations with multiple sites. The right balance among the various communication channels must be struck when communicating and sharing the OEMS vision. Among the preferred approaches are as follows:

- Senior leadership engagement with all levels of the organization – often gathering site workers together to talk with them (during their busy day) thereby implicitly communicating the importance of OEMS
- For multiple-site businesses, a senior executive roadshow can be effective and less costly than bringing people to a central location.
- Print media in the form of OEMS brochures, frameworks, and presentations
- Video presentations of the CEO and senior leaders delivering the OEMS vision messages to other sites

Led by the CEO and Senior Leaders

Despite the various approaches available for communicating and sharing the vision, the most effective approach is personal delivery (face-to-face) by the CEO and senior leaders to all levels of the organization. This method is most effective because it demonstrates the CEO is prepared to take the time from other activities to demonstrate his commitment to OEMS.

8.3 Marketing OEMS throughout the Organization

All Levels of the Organization

There are various schools of thought regarding which levels in the organization the OEMS should be communicated and shared. Some experts advocate only senior leaders and middle managers should be communicated with, since frontline supervisors and workers do not really care. In their view, frontline supervisors and workers are only interested in knowing what has to be done so they can complete their work days.

The authors, on the other hand, advocate all levels of the organization should be communicated with when sharing the vision. The objectives of communication with each stakeholder group are presented in the following table. Key to success in communicating and sharing the vision, however, is changing the messages to ensure the needs and expectations of the various stakeholder groups are met from the communication process.

Level	Communicating and Sharing Objectives
Senior leaders	• Clarification of what is changing and how we get there (the road map) • Provide understanding of alignment with the business strategy • Clarification of change management impact and requirements • Understanding of governance and resources allocation implications
Middle managers	• Understanding of workload and priority changes • Understanding of organization and departmental changes and resources implications
Frontline supervisor and workers	• Understanding of what is changing and how it affects my day-to-day work • Clarification of "What's in it for me" (WIFM)

Find Time to Meet with and Talk with the Frontline Personnel

Finding time to engage with frontline personnel is important in creating a work environment of inclusiveness, and for making all workers a part of the OEMS process and eventual culture shift toward discipline and excellence. For most organizations, frontline workers make up the largest work group and are intimately involved in product and service delivery. Knowing the OEMS vision and accompanying changes helps obtain their commitment and support in the new ways of working.

8.4 Sharpening the Messages – Survey Audience for Implementation Suggestions

Survey Audience after Communication

The authors recommend that for multiple-site organizations, it is important to engage the initial groups communicated with for their input and feedback regarding improved communication in implementing OEMS across the organization.

Some of the suggestions obtained from stakeholder surveys are shown in the following table:

Suggestions	Focus
Elements to focus on	Suggestions to prioritize OEMS elements as follows: 1. Leadership and organizational effectiveness 2. Hazard identification and risk management 3. Working safely 4. Asset reliability and integrity 5. Conduct of operations 6. Training, competency, and human performance 7. Document and information management
Communications and messaging improvement	Communications and messaging improvement suggestions may include the following: • Have a clear OEMS vision • Communicate … communicate … communicate • Involve people – include hourly workers • Managed pace – take it slowly … small bites and continuous improvements • Training – awareness and competency • Leadership commitment – give it time to work • Accountability – hold people accountable
Benefits to communicating more clearly	Some of the potential benefits to communicating more clearly may include: • Informed workforce • Potential efficiency improvements • Safer operations • Standardized and consistent processes • Competent workforce • Best practices application • Improved collaboration, knowledge generation, and sharing • Greater accountability • Improvements in procedure quality and use • Organizational and business performance

8.5 Creating the Change Management Support Organization

Change Management – Skills Required

Transformational leaders are adept at leading and managing change. They are patient and transformational in behaviors, taking the time required to explain why change is required. Change management is a specialized field of expertise and where OEMS is concerned, because of the organizational change impact, a dedicated team of change management experts may be required to support the organization's change management efforts.

The change management organization does not belong to the front line but is adept in the following skills:

- Understanding human behaviors
- Conducting focus group sessions and interviews to understand underlying issues that may impact the change process
- Communications and messaging
- Communicating at all levels of the organization
- Articulating the WIFM
- Writing

Once selected, an effective change management organization is invaluable in enabling OEMS change across the organization.

Managing the Change

OEMS implementation and sustainment depends on an effective change management process and support throughout the implementation process. The authors recommend the 7 Levers of Change Model (Figure 6.8) for greatest effectiveness in change management. Shown in Figure 8.3, the following change management readiness requirements must be addressed to ensure permanent change occurs:

- Developing the change management organization
- Ensuring stakeholder impact assessments are complete, risks mitigated, and gaps closed
- Assigning and managing corrective actions to ensure all gaps are closed in support of effective change
- Developing key messages in support of the change for impacted stakeholders
- Providing recommendations of approach, communication frequency, and medium in support of the proposed change
- Evaluating the effectiveness of approach and adjusting accordingly

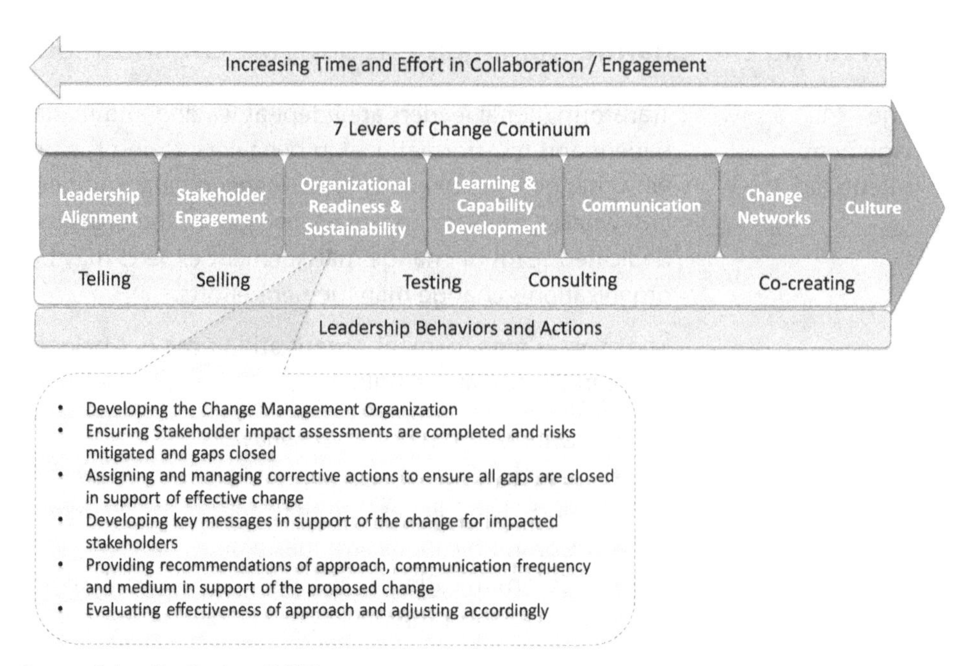

Source: Safety Erudite Inc. (2019).

Figure 8.3: Change management – readiness and sustainability.

8.6 OEMS Information Access

Information Access

To create collective momentum in the pursuit of OEMS, easy access to OEMS information is required by all stakeholders throughout the organization. Where OEMS information is not sensitive and can be shared, the organization should seek to do so providing easy access to information for all workers and applicable stakeholders. Figure 8.4 provides a model for designing and providing electronic access to OEMS information within the organization.

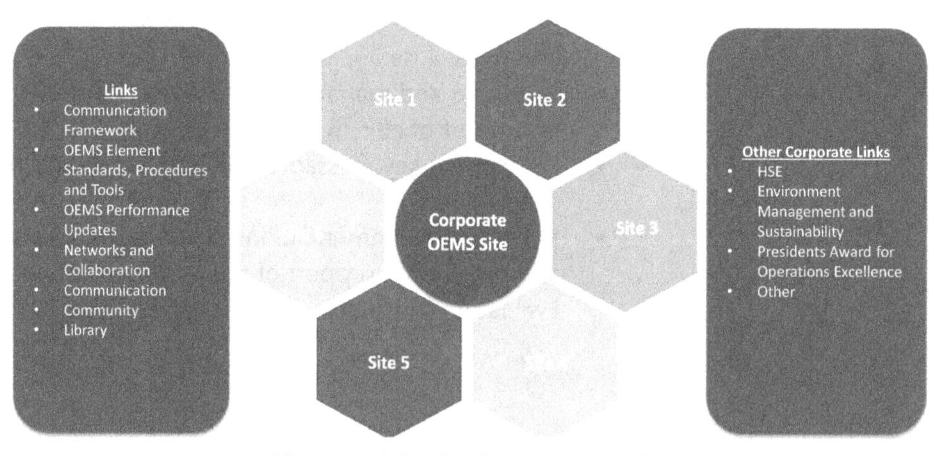

Guidance - Developing the Information Access Architecture

- All Sites are designed similarly and allows for access to each other's OEMS Page
- Minimize clicks to information
- Make it attractive and easy to navigate
- Allow easy return to page
- Make OEMS a central theme

Source: Safety Erudite Inc. (2019).

Figure 8.4: OEMS information access.

8.7 Develop the Implementation Model

Implementation Models

There are several models for implementing OEMS, among which are the following:

- Full implementation across all sites
- Piloting at specific sites
- Implementation of selected elements followed by later implementation of other elements

Pros and cons for each implementation model are provided in the following table:

Implementation Model	Pros	Cons
Full implementation across all sites	• Faster pace of implementation • Effective use of networks in developing consistent approaches to implementation across similar sites	• Can be costly to the organization • Failure to benefit from learnings from piloted implementation
Piloting	• Ability to learn from costly mistakes • Allows more advance sites to move forward and other locations capitalize on the learnings	• Slower implementation • Difficulty in identifying a volunteer organization – *no one wants to be the guinea pig* • Ineffective networks • Site may develop *made here* solutions that may not be applicable to the rest of the organization
Implementation of selected elements	• Less costly and demanding on the organizational resources	• OEMS is an integrated process whereby all elements are integrated – selected implementation fall short of integrated processes

Reference

Safety Erudite Inc. (2019). The Integrated Process Management System Provider. Retrieved June 2019 from www.safetyerudite.com

9 IMPLEMENTATION – FOCUS ON BIG WINS AND KNOWN GAPS

Overview

In almost all organizations, some gaps in the management system are known. Gaps are known based on the following business processes:

- Prior self-assessments and audits relative to the existing management system expectations and requirements
- Corrective actions that may arise from risk management techniques and practices (e.g., process hazard analysis, hazard and operability studies (HAZOP), and layer of protection analysis (LOPA))
- Incident investigations such as failure mode and effects analysis (FMEA), root cause analysis (RCA), and fault tree analysis (FTA)
- Shared learning based on events that may have occurred in other organizations within the industry
- Experienced professionals in the field who have modified work practices to maintain processes and operations without considerations of a management of change (MOC) process and newly introduced risks from modified operations

Figure 9.1 provides an overview of the key sources of known gaps in the management system.

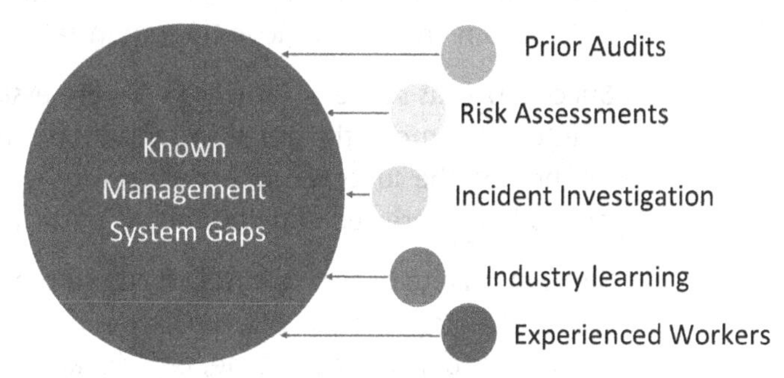

Source: Safety Erudite Inc. (2019).
Figure 9.1: Sources of known gaps in the management system.

9.1 Focus on Big Win Opportunities

What Are Big Wins?

Big wins are the high impact opportunities that can be easily implemented and can deliver the following:

- Reduce operating risk exposures to the business – people, the environment, and assets (in that order)
- Improve the reliability of the assets to prevent unintended downtimes
- Improve worker workloads and work/life balance
- Low-cost, high-value opportunities
- High risk-reduction, low-complexity opportunities

Demonstrate Commitment to Workers

Among the biggest win opportunities are those that are designed to reduce the workloads of workers without creating concerns of job loss and retrenchments. These types of big wins help create support for the management system and are very instrumental in creating ownership from the ground up. What is most important to frontline workers is often buried in the *What's in it for me* (WIFM) proposal. The bigger the prize to the worker, the greater the buy-in.

In most instances, workers are looking for things that can

- Improve the safety of all workers, particularly themselves
- Reduce workloads associated with unplanned outages, which generally leads to hectic work days and overloads
- Simplify activities that allow a better work experience

These are generally activities that do not require immense amounts of planning to get them moving forward. However, satisfying these needs creates tremendous support and momentum in moving the implementation of the management system forward.

Why Big Win Opportunities

Big win opportunities are visible and start the process. People are motivated when they see movement in the process, and big win opportunities create the momentum required to implement a management system. Success in the implementation of any management system depends heavily on the collective momentum created to move the organization forward.

Success breeds success. Early big wins are essential to create a groundswell of interest to move the entire organization forward. Communicating and celebrating the successes of early big wins is instrumental in building the internal network of interested and committed participants.

Similarly, failure breeds failure. It is, therefore, imperative to get it right the first time on each of the big wins attempted. Plan carefully and commit the right levels of resources – time, people, and dollars – to ensure success. Optics (perception) are important to avoid apathy and general discord associated with failed attempts.

9.2 Prioritizing Gaps and Opportunities

Overview

In most organizations seeking to implement an OEMS, attempting to close all identified gaps simultaneously places an enormous squeeze on the organizational budget. In addition, there may be insufficient trained and competent personnel internal to the organization to undertake the task.

Prioritizing gaps for closure becomes an important task in ensuring capital and labor load leveling for sustainable implementation.

Use of the Priority Matrix

As explained earlier in Chapter 2, the priority matrix is among the most effective processes for determining how gaps shall be closed in a prioritized manner.
In this exercise, a preliminary review of each gap is undertaken to assess the following:

- The complexity and effort required in closing the gap
- The risk reduction and potential value generated from closing the gap

Once these are known for all known gaps, they are plotted unto the priority matrix as shown in Figure 9.2. It provides the organization a reasonable barometer regarding order in which gaps should be closed. The closer we are to zero, the higher the priority.

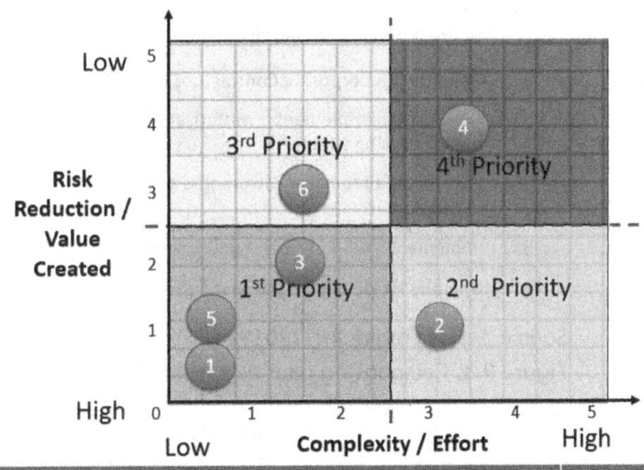

	Gap Closure Activity – Known Gaps	Risk Reduction / Value Created	Complexity / Effort
1	Upgrade and implement the upgraded MOC Process	0.5	1.0
2	Upgrade the Corporate Contractor Safety Management System	1.5	3.5
3	Develop and implement a winterization program to limit unplanned outages during winter	2.0	2.0
4	Implement a regulatory awareness program to be on the leading edge of new regulations	4.0	4.0
5	Implement a working alone policy for remote location workers	1.0	1.0
6	Upgrade Incident Management software and train personnel	4.0	2.0

Source: Safety Erudite Inc. (2019).
Figure 9.2: Priority matrix for prioritizing known management systems gaps.

Categorizing Gaps

The other, less applied, method used for prioritizing known gaps is the approach used in categorizing them as shown in Figure 9.3. Gaps are categorized into the following categories for prioritization:

- At-risk gaps
- Low-hanging fruits
- Best practices

Generally, at-risk gaps and low-hanging fruits will be closed in priority to other gaps. This method is more subjective but generally leads to a similar outcome as that of the priority matrix.

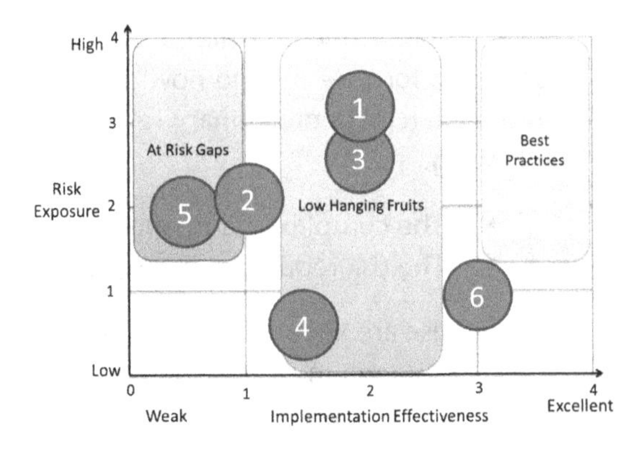

	Gap Closure Activity – Known Gaps	Risk Exposure	Implementation Effectiveness
1	Upgrade and implement the upgraded MOC Process	3.5	2.0
2	Upgrade the Corporate Contractor Safety Management System	2.0	1.0
3	Develop and implement a winterization program to limit unplanned outages during winter	2.5	2.0
4	Implement a regulatory awareness program to be on the leading edge of new regulations	0.5	0.5
5	Implement a working alone policy for remote location workers	2.0	0.5
6	Upgrade Incident Management software and train personnel	1.0	1.0

Source: Safety Erudite Inc. (2019).

Figure 9.3: Categorizing gaps as a means for prioritizing them.

9.3 Develop Consistent and Standardized Solutions

Overview

For many organizations with multiple business units, there is a tendency for business unit leaders and/or facility leaders to pursue solutions on their own. Thus, for the same situation the organization ends up with multiple ways of resolving the issue. Indeed, this results in a high-cost solution as well as a low-cost solution.

The authors recommend creating an environment where leaders work together toward consistent solutions for common gaps when they exist in the organization's management system. This is where the value of networks becomes apparent to the business. Figure 9.4 demonstrates the process for generating common and consistent solutions for common gaps identified across the business.

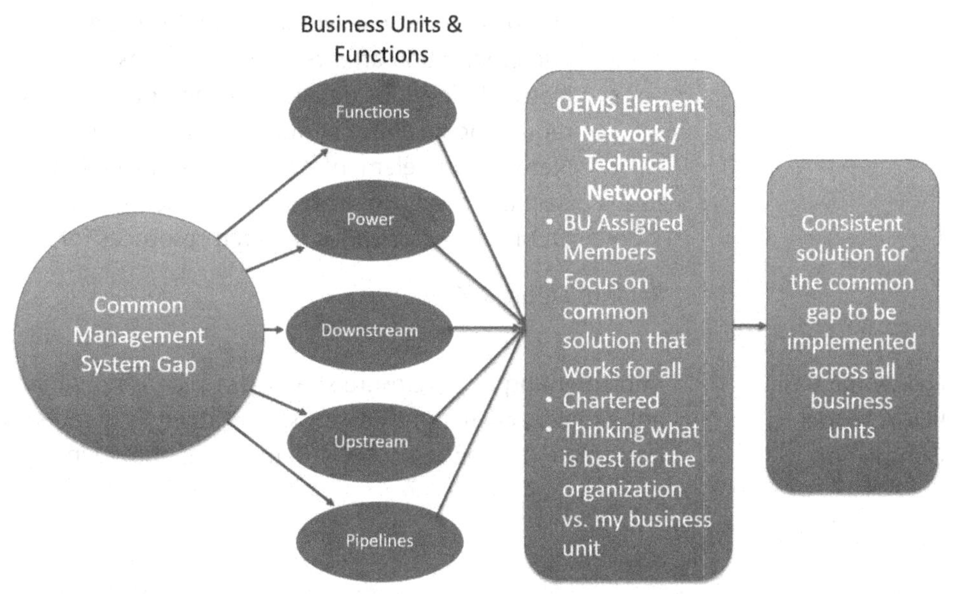

Source: Safety Erudite Inc. (2019).
Figure 9.4: Consistent solutions for common gaps – leveraging networks.

Common Management System Gaps

Regardless of the business unit and functional group, there are many common management system gaps that are likely to surface. The following table provides an example of common gaps that can benefit from common or consistent solutions across the organization:

Gap	Common	BU/Function-Specific
Upgrade and implement the upgraded MOC process	✓	
Upgrade the corporate contractor safety management system	✓	
Develop and implement a winterization program to limit unplanned outages during winter		✓
Implement a regulatory awareness program to be on the leading edge of new regulations	✓	
Implement a working alone policy for remote location workers		✓
Upgrade incident management software and train personnel	✓	

9.4 Develop Implementation Plan for Each Prioritized Gap

Overview

For each gap closure strategy, the authors recommend following a disciplined process regarding implementation. As discussed earlier in Chapter 2, implementation of every gap closure strategy should be as follows:

- Network develops gap closure solution
- Input provided by internal/external subject matter experts and community of practice

- Network leader receives approval from element owner to proceed with implementation after reviewing options available
- Network develops detailed implementation inclusive of resources requirements and schedule for completion
- Network and element owner present to OEMS stewardship team and gain approval to proceed with implementation
- OEMS owner provides required resources for implementation

Refer to Figure 2.6.

Follow the Basic PDCA Model to Completion

Implementing a gap closure strategy or plan is no different from implementing any project activity. In all cases, sustainability is greatest when we follow the basic plan–do–check–act (PDCA) model as shown in Figure 9.5.

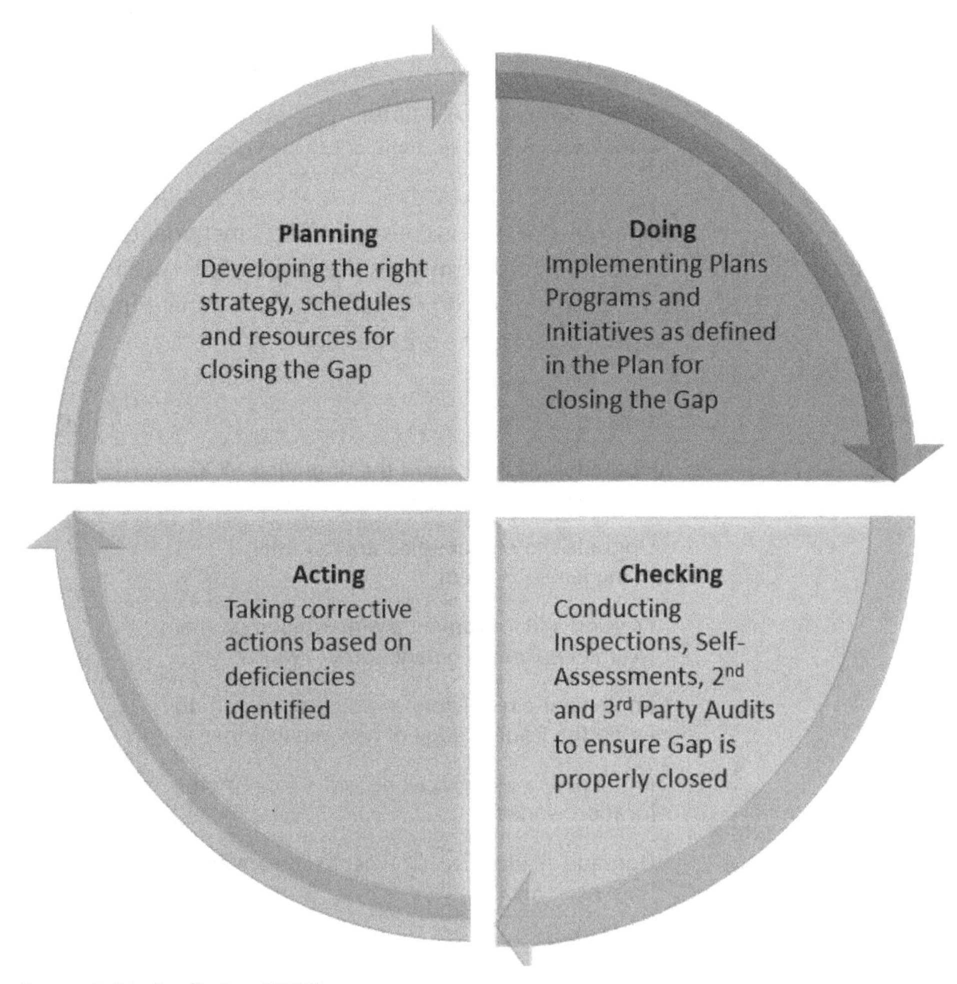

Source: Safety Erudite Inc. (2019).

Figure 9.5: PDCA model for managing the gap closure activities.

Reference

Safety Erudite Inc. (2019). The Integrated Process Management System Provider. Retrieved June 2019 from www.safetyerudite.com

10 BUSINESS UNIT (BU)/FUNCTION IMPLEMENTATION – FOCUSING ON MUST-HAVES AND KNOWN GAPS

Overview

The implementation phase of OEMS is perhaps the most difficult for any organization. The magnitude of the effort has only just revealed itself to the organizational leaders and most often, the response by the impulsive leaders is *let's not touch this monster*. For the committed leader, however, the response is *let's find a way to make this work*. Recognizing OEMS is a >8-year journey; the approach has to be slow and steady.

In this chapter, the authors provide a systematic approach to taking the implementation of OEMS forward in a measured way. The approach requires setting the organization up for success by identifying and fixing the *must-haves* and *known gaps* identified by the business in order to sustain current levels of operation and business in a safe and reliable manner.

What Are *Must-Haves*?

Must-haves are fundamental requirements and processes essential for safe and reliable operations of the business. In a process operating environment, among the *must-haves* may include the following for process operations:

- Leadership development – with focus on
  - High-reliability behaviors
  - ABCD of trust
  - Improving leadership visibility
  - Transformational leadership behaviors – inspiring hearts and minds
- Change management – people and organizational changes
- Performance metrics consolidation and prioritization
- Electronic logs development – consistent log requirements across all units/sites
- Documentation management storage and access management – mimic at sites
- Operating envelopes – safe operating limits defined, monitored, and exceedances actioned
- Seasonal planning – training roles of maintenance and operations
- A competent workforce
  - Operator development and competency-based training
  - Walk the line – detailed site training for all operations organizations
  - Operator-driven reliability – operator training
- Continuous improvements – networks – training, setup, and collaboration
  - Shared learning process – electronic and simplified
- Risk management processes including risk registers, risk matrix, and risk management tools such as Process Hazard Analysis (PHA) and management of Change (MOC).

What Are *Known Gaps*?

Known gaps, on the other hand, include activities, requirements, and processes that are generally known by the organization but have been avoided. Cost and schedule constraints, insufficient effort toward addressment, and the perception that these gaps are not impactful, all contribute to the prevalence of known gaps. Known gaps that have not been addressed over an extended period illustrate to followers that these deviations are acceptable and constitute normalcy. Known gaps may include any of the following:

- Any must-haves identified earlier
- May be simple basic needs such as portable toilets to capital items such as critical pieces of equipment or assets

Why to Address Must-Haves and Known Gaps First

Addressing must-haves and known gaps provide tremendous advantage to the business. Benefits for doing so are detailed in the following table:

	Benefits
Must-haves	• Sets the foundation for a more successful implementation • Helps manage the cost and pace of implementation • Addresses high-risk prioritized • opportunities first • Signals to the business its seriousness about implementing OEMS • Learning opportunities from early implementation
Known gaps	• In addition to the benefits provided for must-haves, the credibility of the organization is enhanced

Furthermore, leader credibility would be severely eroded should the organization begin undertaking self-assessments and second- and third-party audits to identify gaps before the must-haves and known gaps are addressed.

Identifying Gaps and Must-Haves

Gaps and must-haves are identified as follows:

- Using a systematic process of interviews of senior leaders at each site/facility
- Asking the questions, "what keeps you up at night?"
- Objective reviews of 2–3 years' prior audits and surveys conducted across each site/facility

Figure 10.1 provides an overview of the process used for identifying gaps.

- Independent interview of Site Leader / Selected Department Manager Element Owner / Network Member across each Site/Facility
- Review of prior 2-years audits findings for each Site/Facility

Interviewee	Focus
Site / Facility Leader	All Elements
Selected Department Managers	All Elements
OEMS Element Owner	Assigned Element
Element Network Member	Assigned Element
Audit Findings Reviews (2 Years Prior)	
Audit Reviews (2 Years Prior to Implementation)	OEMS / Culture
	HSE
	PSM

Source: Safety Erudite Inc. (2019).

Figure 10.1: Process for identifying must-haves and known gaps.

10.1 Classifying Gaps into Strategic and Site/Facility-Specific

Classifying Gaps – Must-Haves and Known Gaps

For many organizations, gaps – must-haves and known can be classified into two categories as follows:

- Strategic gaps
- Site/facility-specific gaps

Each type of gap is addressed differently in the planning and implementation stages. Figure 10.2 provides an overview of the ways each type of gap is managed.

√ - Strategic Gaps	• Gap Closure Plans developed by the Element Network – Single Solution • Implementation plans approved by the Corporate Leadership Team • Prioritization, resourcing and schedules for implementation defined by the Corporate Leadership Team • Implemented at the Site by the Element Network Member
√ - Site Specific Gaps	• Developed by the Site/Facility • Approved by the Site/facility Leadership Team / Site Leader • Prioritization, resourcing and schedules for implementation defined by the Site Leader • Plans implemented by the Site/Facility Network Member ***Note:*** *A Site/facility may consider full scale adoption from another Site/Facility to reduce workload and ensure consistency*

Source: Safety Erudite Inc. (2019).

Figure 10.2: Strategic and site/facility-specific gap management.

BU Strategic and Site/ Facility-Specific Gaps

Understanding the amount of work required by the Business Unit (BU) and its site/facility is important to the business in prioritizing and resourcing efforts. Using the process earlier discussed to identify and classify must-haves and known gaps, the matrix provided in Figure 10.3 provides an overview of the sample scope of work required in addressing these two buckets of work at the BU and site/facility levels.

√	Strategic Gaps are found across all Sites/Facilities – A single solution may work for correcting the gap across all sites
√	Site/Facility-specific Gaps may be limited to a specific or few Site/Facility. Solutions may be borrowed and applied from other Sites/Facilities

Strategic Gaps / Self Assessments /2nd & 3rd Party Audits	Site 1	Site 2	Site 3	Site 4	Site 5	Site 6
Leadership Development	√	√	√	√	√	√
Competent Workforce	√	√	√	√	√	√
Walk The Line		√		√	√	
Operating Envelopes	√	√	√	√	√	√
Winterization Program			√		√	
Documentation		√	√			
Networks – Training, Setup and collaboration	√	√	√	√	√	√
Shared Learning Process	√	√	√	√	√	√

Source: Safety Erudite Inc. (2019).

Figure 10.3: Mapping the scope of work across the BU and sites/facilities.

10.2 Scheduling the Work to Close These Gaps

Scheduling Process

Once the must-haves and known gaps have been established at both the corporate and BU levels, a schedule should be developed to track and manage the implementation. Figure 10.4 provides a high-level overview of the major activities and schedule for implementation of OEMS across an organization.

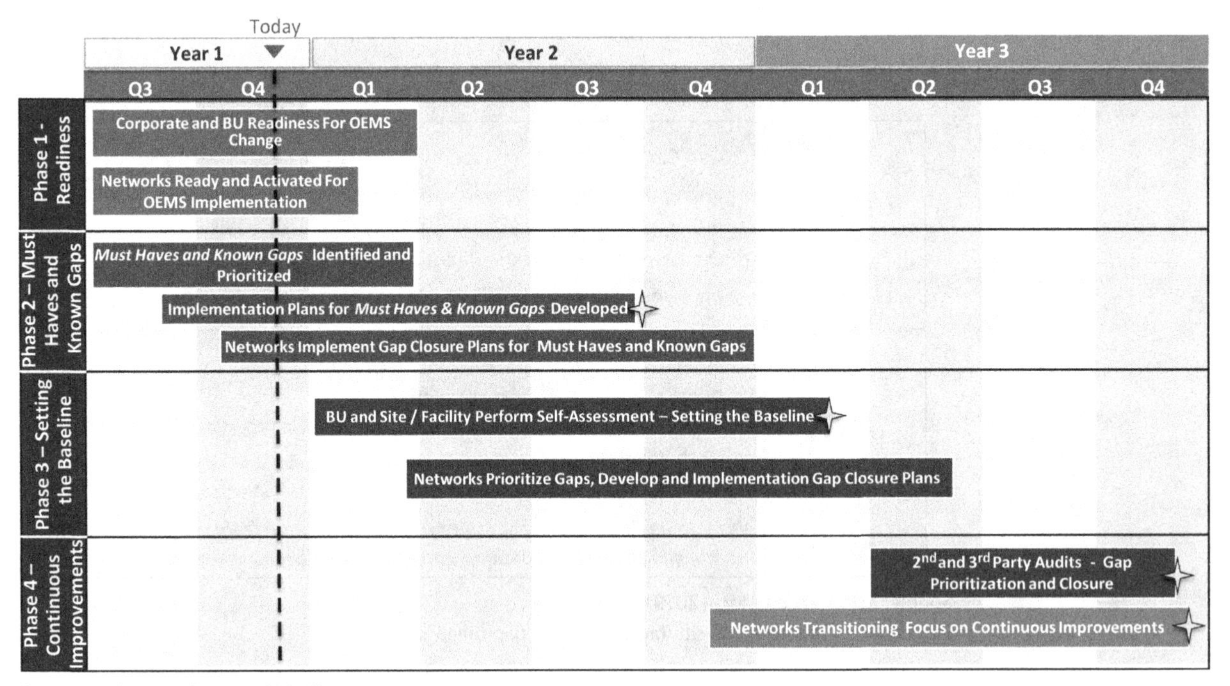

Source: Safety Erudite Inc. (2019).

Figure 10.4: High-level major activities and schedule for implementing OEMS.

Care must be applied in determining the right pace and schedule for implementation based on the following:

- Number of must-haves and known gaps to be closed
- Resourcing capabilities of the organization

- Human resources impact – workloads and burnout concerns
- Organizational appetite for change
- Amount of organizational change taking place simultaneously

In all instances, handled properly with adequate application of change management principles, sustained transition to operations excellence is achievable.

10.3 Establishing the Baseline

Overview

Once the must-haves and known gaps have been addressed, the organization should establish the OEMS baseline at the following levels:

- Corporate
- BU
- Site/facility

Knowing the baseline at each level allows the organization to perform internal benchmarking on its OEMS progress. The benchmark provides an excellent measure for year-over-year progress relative to requirements of the management system and places accountability on element-related personnel for annual progress and achievements.

Establishing the Baseline – Self-Assessment

The baseline can be achieved through several methods, all of which depend on a comparison between requirements detailed in the OEMS standards relative to the actual status of the requirement at each of the corporate, BU, and site/facility levels.

In all cases, the baseline is determined by a systematic audit process designed to measure how well the requirement has been met by the organization, the BU, and the site/facility. The status of the requirement is compared relative to the following measures:

- Leadership commitment
- How well the risk exposures are being managed
- Are the desired performance levels being derived?
- How well the requirement is being implemented

A score is allocated for each of these measures consistent with the OEMS maturity map for the organization.

Scoring is rolled up to provide maturity scores as follows (refer to Figures 10.5–10.8):

- Sub-element
- Element
- Management system – site/facility
- Management system – BU
- Management system – organization

- Review the status of each requirement / sub-element relative to the measures that include Leadership Commitment; How well the risk exposures are being managed; Are the desired performance levels being derived; How well the requirement is being implement
- Assess against the Corporate OEMS Maturity Map
- Allocate score accordingly
- Identify gap/constraints in achieving a score of 3.0 on the OEMS Maturity Map

Element	***Sub –Element / Requirements***	***Gaps Identified***	***Avg. Score***
Management of Change	• A formal MOC Process must exist for managing change ○ A MOC tool is available and implemented across all Sites/Facilities ○ Requirement ***n*** • Change is authorized • A stakeholder impact assessment is completed and risks mitigated • Change is communicated • Changes is closed out • Sub-Element ***n***	• Standardized tools for managing change not available • Individuals identified vs. roles for assigned responsibilities • Inconsistent signoff of all required authorities for approving change • Very effective use of email, alerts, communication and training tools for communicating change • Inconsistent approached to managing the effectiveness of change. Documentation absent	2.0 2.0 2.5 3.5 1.5
Element Score			**2.3**

Source: Safety Erudite Inc. (2019).

Figure 10.5: Maturity scoring of OEMS element requirement.

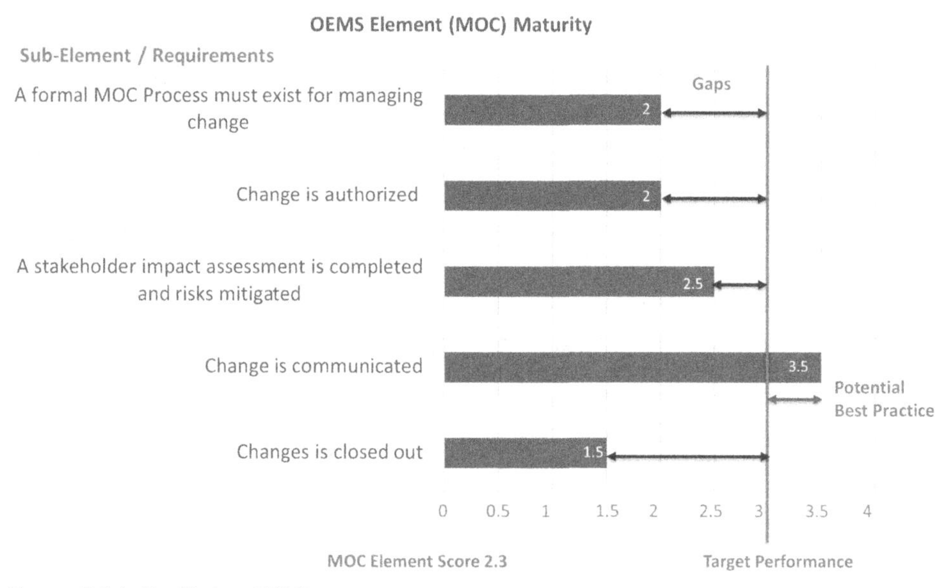

Source: Safety Erudite Inc. (2019).

Figure 10.6: Maturity scoring of OEMS element – MoC.

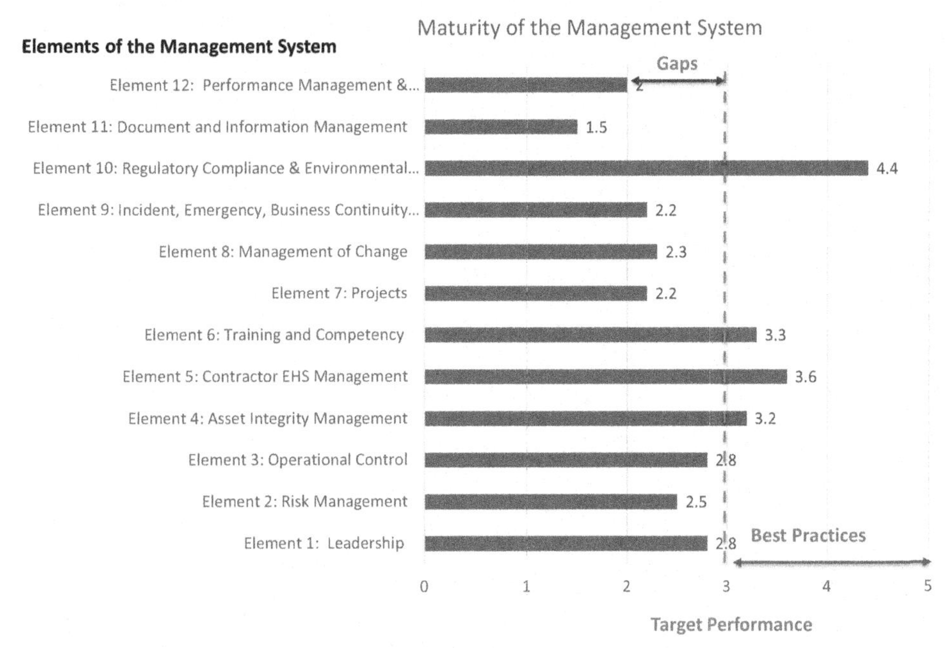

Source: Safety Erudite Inc. (2019).

Figure 10.7: Self-assessment – OEMS baseline for each element.

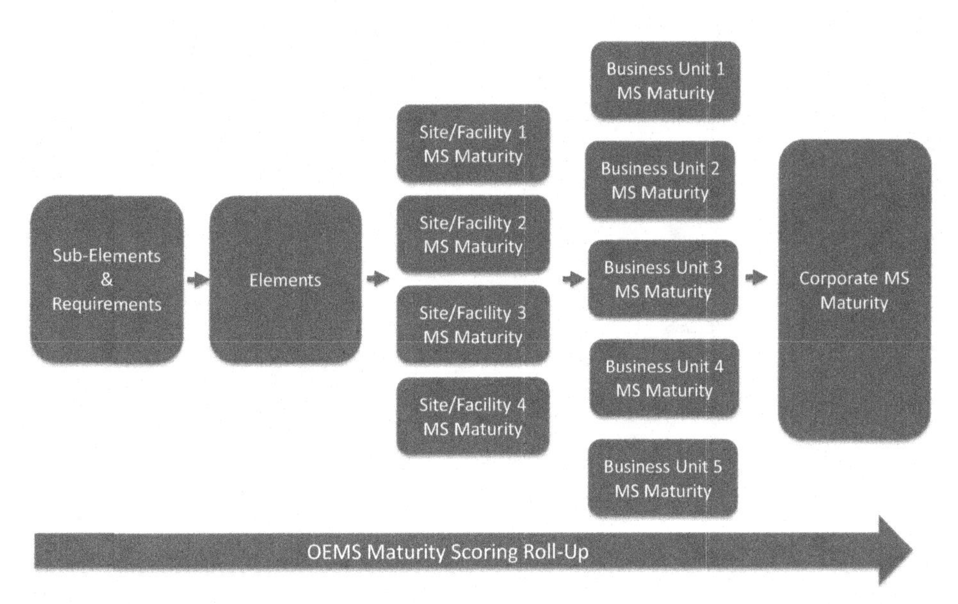

Source: Safety Erudite Inc. (2019.

Figure 10.8: Maturity scoring of OEMS element, site/facility, BU, and corporation.

Reference

Safety Erudite Inc. (2019). The Integrated Process Management System Provider. Retrieved June 2019 from www.safetyerudite.com

11 AUDITING FOR CONFORMANCE AND COMPLIANCE

Objective of Audits

Auditing of the management system is an ongoing process that is designed to achieve the following:

- Proactively identify and close gaps in the management system before loss, incidents, or suboptimal performance.
- Enhance business performance, providing information to leaders regarding risk management opportunities.
- Benchmark and evaluate the organization's level of maturity that is essential in decision-making and business improvement.
- Define the level of conformance and compliance to the organizational requirements.

Auditing for conformance and compliance is best achieved by second- and third-party audits. In this section, the authors provide an overview of the management system's audit process for achieving the best audit performance.

Challenges in Auditing

Challenges in auditing include the following:

- Variability in auditors' analysis and scoring of the same scenario
- Sites are often resource-constrained in performing and supporting detailed audits
- Roncea (2016) advised on the following challenges regarding audit report:
    - "Use standard expressions difficult to understand for those who do not know in detail the reference standards
    - Deal with compliance rather than effectiveness
    - Contain only a few recommendations for improvement and generally, do not refer to solutions/measures which can be taken for effectiveness improvement" (p. 786)

Eliminating Auditors' Inconsistencies

Eliminating inconsistencies in auditors' assessments is achieved through the following methods:

- Developing requirements interpretation guidelines and terminology that are consistently used by all auditors
- Use of an automated audit process that eliminates personal biases and individual judgment. Rather, the audit process is linked to the maturity map for scoring, gap identification, and Procedure, Accountability, Competency and Assurance (PACA) Model for determining more accurate corrective actions
- Well-trained and competent auditors and audit organizations
- Simple audit tools to enable consistency in the audit process such as
    - A defined maturity map
    - PACA-related suggested corrective actions

- Audit report templates
- Audit protocols designed to ask the right questions and generate accurate and transparent responses from auditees

The Audit Process

The audit process is a well-planned event that is known well in advance by the site or business unit (BU). An annual audit plan should be developed and communicated at the start of the physical year. Sites and BUs are formally notified of the scope of the audit and are provided adequate time and opportunity to be ready for the event. It is critical to schedule audits in coordination with the sites and BU to prevent conflicts with local activities.

The audit process generally provides the following:

- A lead auditor and a site/BU coordinator
- Schedule of the event
- Auditors and auditees
- Reviews of prior audit reports and findings
- Pre-read information provided by the site/BU
- Audit protocols
- Minimal disruptions to normal operations of the business

Major Audit Activities

Major activities involved in audit preparation and planning, the audit, and post-completion of the audit are provided in Figure 11.1.

Pre-Audit Planning	Audit	Post Audit Activities
• Audit Scope defined • Audit team identified • Audit Lead and Site / BU audit Coordinator assigned • Audit schedule defined • Pre-read materials provided • Auditors and auditees begin the engagement process and prepare for the audit	• Audit kickoff meeting –Audit executed as per schedule • Daily debrief with Site / BU leader performed – High risk gaps to be communicated immediately to the Site / BU leader • Audit closing - Site / BU leader and all auditees • Provide high level summary of finding and gaps to all participants	• Audit report preparation • Gaps defined and assigned a risk ranking • Site / BU leader to prioritize gaps, define and assign corrective actions • Site / BU leaders to resource, implement and steward corrective actions • A verifier confirms that corrective actions are adequately implemented to adequately close gaps identified during the audit • Audit Finalized and closed

Source: Safety Erudite Inc. (2019).

Figure 11.1: Major audit activities.

High-Level Audit Process

A schematic of the audit process is shown in Figure 11.2.

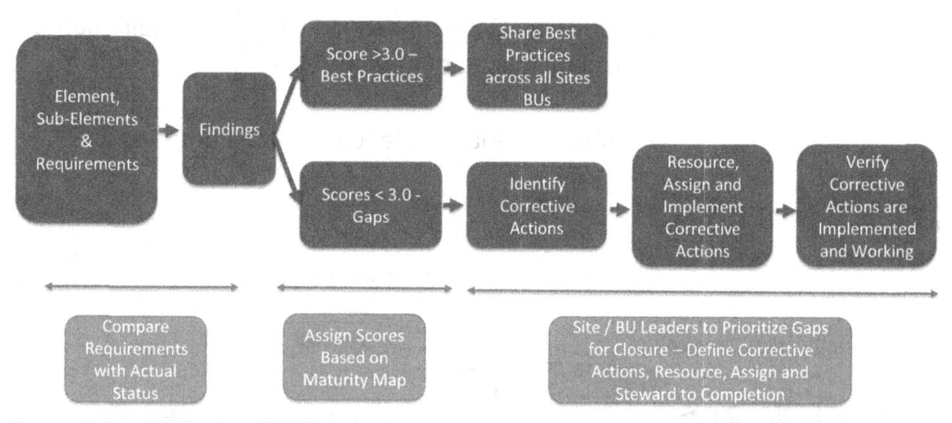

Source: Safety Erudite Inc. (2019).
Figure 11.2: High-level audit process.

Auditing – Allocating Maturity Scores

The auditing process requires a disciplined approach to the process as demonstrated in Figure 11.3. The process involves a comparison between the requirements of the management system and how well the following is demonstrated:

- Leadership commitment
- Risk management principles
- Desired performance outcomes
- Process implementation

	REGRESSIVE	REACTIVE	PLANNED	PROACTIVE	EXCELLENCE
Leadership	Leaders are disengaged with little knowledge and appreciation of MS, its goals and objectives	Partial knowledge of MS expectations Demonstrates direct leadership only when required within MS Treat MS as a nice to have process Not fully bought in into the process	MS forms the basis for decision making Demonstrates consistent visible, active leadership, commitment and engagement (Resources, Continual Improvement, Ownership, Driver of MS)	Ownership of MS throughout aspects of planning, implementation, checking and review of their MS Focused on leading change and continuous improvements Leaders are able to make effective decisions based on outputs of MS to pursue continual improvement	Full commitment to MS A champion internally and externally Actively initiates and participates in MS improvements and is innovative in implementing effectively Continually drives accountability in management system throughout all levels
Risk Management	Basis for decisions are not clear Decisions do not clearly take into account overall view of organizational risks Risk –based decisions are not documented	Decisions do not reflect level of various organizational risks identified Some use of the risk matrix in decision-making	Strong understanding and application of risk-based decision-making Decisions are taken and documented based on the level of risk identified throughout organization Risks are mitigated to ALARP levels	Organizational decisions consistently include evaluation of risks as an input. Review mechanisms in place to verify decisions are taking risk into account.	Risk management information fully integrated into Business Planning processes and organizational culture
Performance	Performance rarely or not assessed Metrics are not available and a general sense of firefighting and high stress exists Performance not understood	Ad-hoc / inconsistent review of performance Performance improvements strategies not clear Prepared to continue as is	Performance evaluation process established and implemented Performance measured against a pre-defined comprehensive set of performance indicators relative to subject of evaluation KPIs are available and stewarded	Outputs of performance evaluation drives continual improvement Feedback process implemented to inform relevant stakeholders of results Leading indicators being used for timely intervention and actions	Based on improved performance, the Organization realizes overall improvement When compared with industry (benchmarking), Performance is top quartile of industry
Process Implementation	Very few or no process developed Heavy reliance on people – a people dependent process No Standards or procedures to support the work- People dependent	Process developed but not fully aligned with key components Processes do not include the expected level of accountability /responsibility to be effective Processes developed but not effectively implemented	Processes established (key components, resources, accountabilities, responsibilities) and implemented to meet requirements	Integrated MS Processes effectively linking to other relevant MS Elements / Business Processes Measurement of process effectiveness being used to initiate continual improvements Benchmarking of processes across the Site	Benchmarking of processes across organization and/or industry Process is recognized as a best practice across the company and/or industry
Score	0.25	0.5	0.75	1.0	1.25

Source: Safety Erudite Inc. (2019).
Figure 11.3: Audit process – determining requirement maturity score.

Maturity of each requirement is cross-referenced to the management system maturity map shown in Figure 11.4, which is repeated here for your convenience. Gaps are defined when the requirement score is <3.0 on the management system maturity map.

Organizational Management System Maturity Model

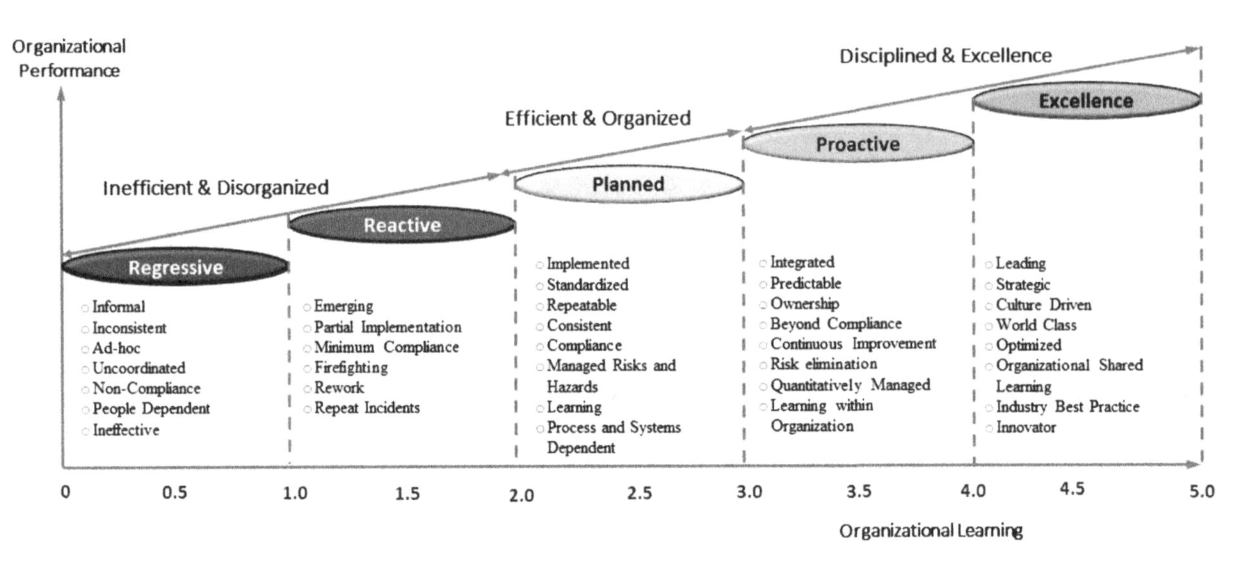

- Applicable Sources defined by level of compliance to the element requirements
- At the very minimum, the Organization must strive for planned compliance levels

Source: Safety Erudite Inc. (2019).

Figure 11.4: Management system maturity map.

Gaps in the Management System

Gaps in the management system are generally classified as follows:

- Single-point gaps – A gap that has resulted from a missed action. Once corrected, the gap is eliminated in a sustained way.
- Systematic gaps – Gaps resulting from management system deficiencies and require a coordinated and systematic approach to sustainably fixing the gap.

In all cases, gaps must be closed by corrective actions that are sustainable.

11.1 Corrective Actions – Following the PACA Principles

Corrective Actions – How We Define Them

Corrective actions are generally suggested by the auditor during the audit process. They are further refined by the site/BU leader to be more appropriate and specific for the business. In most instances, corrective actions provided by auditors are high level and often leaves site/BU leaders confused and uncertain regarding what the appropriate corrective actions are. PACA principles and model provide a process for developing corrective actions that are specific and sustainable.

What Are PACA Principles?

PACA principles offer a systematic approach to identifying sustainable corrective actions identified in the management system. In almost all cases, when gaps are identified in the management system, corrective actions almost always fall into one or more of the following categories:

- Procedure deficiencies
- Accountability issues
- Competency deficiencies
- Assurance issues

PACA principles require the auditor to systematically address each PACA category to identify corrective actions associated with each gap identified in the management system to develop sustainable corrective actions.

Figure 11.5 provides an overview of the PACA model for determining corrective actions associated with gaps identified during the audit.

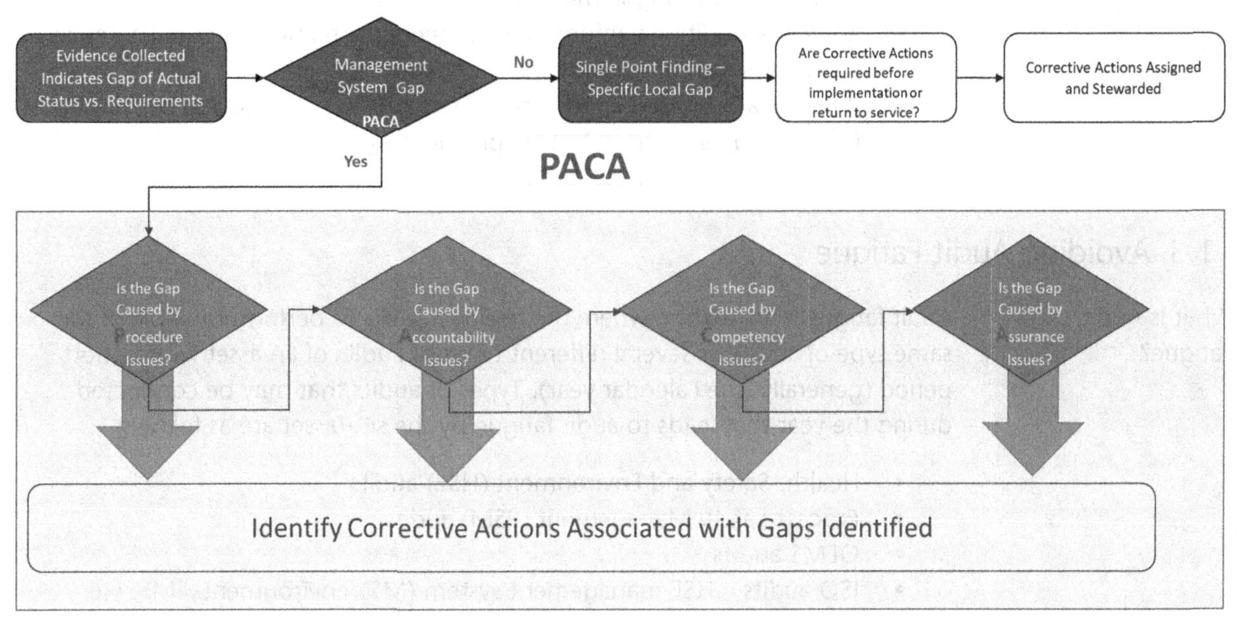

Source: Safety Erudite Inc. (2019).
Figure 11.5: PACA model for sustainable corrective actions.

Corrective Action Management – Robust Process

Ensuring a robust corrective action management process exists for closing gaps identified during the audit is essential for continuous improvements in the management system. Attributes of a robust corrective action management process include the following:

- Adequate resourcing
- Assigned deliverables and accountabilities
- Start and finish dates
- Visibility on progress – easily accessed status updates
- Reminders to both assigned workers and supervisors so that corrective actions are not overlooked
- An online process that is easy to use

11.2 Sharing Best Practices

Sharing Best Practices

Sharing of best practices is an often overlooked aspect of the audit process. Once identified, best practices should be easily shared across similar sites so that collective organizational benefits are derived. This is particularly important when the business operates multiple sites that produce essentially the same products using similar technology and practices. Shared best practices allow all sites to do the same thing the same way, thereby optimizing performance.

Attributes of Good Sharing

Good sharing of best practices is the outcome of a commitment among all sites to share learning and best practices in an organized way. The authors recommend online processes that prevent dilution of the value of the best practice or learning as it is transmitted from one site to others. Attributes of a good shared learning process include the following:

- Consistent and simple messages
- Source of additional information, if needed – name and contact details of someone who can provide more details
- Provides a push/pull approach to sharing of knowledge and best practices
- Online and easy access to best practices

11.3 Avoiding Audit Fatigue

What Is Audit Fatigue?

Audit fatigue is a situation where the business seeks to perform multiple of the same type of audits or several different types of audits of an asset within short period (generally one calendar year). Types of audits that may be conducted during the year that leads to audit fatigue by the site/asset are as follows:

- Health, Safety and Environment (HSE) audits
- Process Safety Management (PSM) audits
- OEMS audits
- ISO audits – HSE management system (MS)/environmental MS, etc.

In such instances, the site/asset is continually preparing for audits and often does not have the time to implement corrective actions from the prior audit.

Avoiding Audit Fatigue

Preventing audit fatigue is essential for maintaining measured progress toward excellence by the business. In many instances, it's easy to combine HSE/PSM and OEMS into a single audit with the right probing questions defined in the audit protocols of the business.

In this way, a single audit meets the requirements of HSE/PSM and OEMS. The auditing objectives of proactive gap identification and application of corrective actions are enhanced with the following:

- The application of well-defined element standards and requirements
- Flexible audit protocols that enable the auditors to be probed in a noninvasive way
- The application of PACA to deliver specific corrective actions to gaps identified

By applying a single audit (generally OEMS) that covers HSE and PSM requirements, the business avoids audit fatigue and develops a solid baseline from which performance improvements can be generated in a sustainable manner.

Recommended Audit Frequency

The following are recommended frequencies of audits for OEMS:

Audit Type	Recommended Frequency
• Self-assessment	• Self-assessments are recommended annually.
• 2nd -party audits	• 2nd party audits should be performed on average every 2 years based on the performance and progress of the site/asset/BU.
• 3rd -party audits	• 3rd -party audits are recommended on a 3-year cycle and should be implemented only to validate the findings and gaps identified by second-party audits or when major incidents are chronic to the business.

11.4 Preparing the Audit Report

What Are Audit Reports?

Audit reports are designed to achieve the following:

- Document findings, gaps, and corrective actions by the site/asset/BU when an audit is completed.
- Communicate findings, gaps, and corrective actions to senior leaders of the organization.
- Identify work priorities of the business that may be applied in defining the annual budgets of the business.

Layout of the Audit Report

An audit report should contain the following sections:

- An executive summary
- Detailed assessment of each element of the management system that is audited

Executive Summary

The executive summary is designed to communicate the site/asset/BU conformance to the management system requirements to senior leaders of the organization. Figures 11.6 and 11.7 provide an overview of the content of the executive summary.

<table>
<tr><td colspan="5">Executive Summary of Audit</td></tr>
<tr><td colspan="5"><u>Summary Comments</u>
An OI Assessment was conducted at the named Site with a focus on the following Elements:
• List OEMS Elements Assessed (Audited)

The assessment team identified the following findings requiring corrective actions:</td></tr>
<tr><td rowspan="2">Finding (non-conformances)</td><td rowspan="2">Total</td><td colspan="3">Risk Ranking</td></tr>
<tr><td>Critical</td><td>Serious</td><td>Moderate</td></tr>
<tr><td>Single Point
• XX
• XX</td><td></td><td></td><td></td><td></td></tr>
<tr><td>Systematic – Management System
• XX
• XX</td><td></td><td></td><td></td><td></td></tr>
<tr><td>Total</td><td></td><td></td><td></td><td></td></tr>
<tr><td colspan="5">1. Provide Graphic Summary of Element relative to maturity map</td></tr>
</table>

Lead Auditor	All Auditors / Specialists have contributed to, reviewed and accepted this report	Date
<as appropriate>	<as appropriate>	dd-mmm-yyyy

Source: Safety Erudite Inc. (2019).

Figure 11.6: Executive summary of audit.

A graphic summary of the maturity of each element is provided in the radar diagram shown in Figure 11.6.

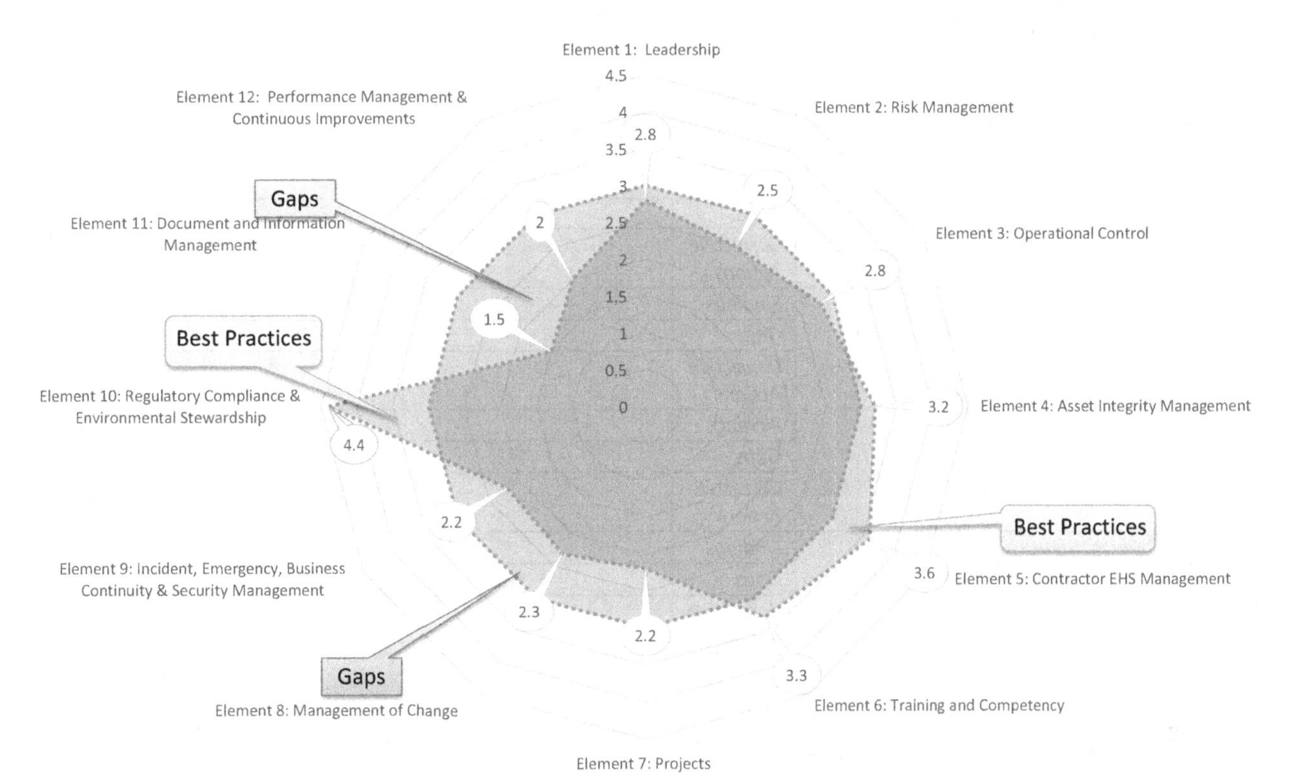

Source: Safety Erudite Inc. (2019).

Figure 11.7: Summary management system maturity – radar diagram.

Detailed Assessment

The detailed assessment section of the audit report is designed to communicate the following to site/asset/BU leaders:

- Findings relative to requirements of the element
- Gaps in the management system relative to the requirements of the element
- Approved corrective actions consistent with PACA principles

Figure 11.8 provides a suggested layout for communicating detailed components of the audit report.

Detailed Audit Report - Elements

	Element: 1			
#	**Requirements**	**Risk**		
1	Entered in the order of requirements in the OI Element		*Findings*	
			Gaps	
			Corrective Actions	
2	Requirement 2		*Findings*	
			Gaps	
			Corrective Actions	
3	Requirement 3		*Findings*	
			Gaps	
			Corrective Actions	
4	Requirement 4		*Findings*	
			Gaps	
			Corrective Actions	
5	Requirement 5		*Findings*	
			Gaps	
			Corrective Actions	

	Element: 2			
#	**Requirements**	**Risk**		
1	Entered in the order of requirements in the OI Element		*Findings*	
			Gaps	
			Corrective Actions	
2	Requirement 2		*Findings*	
			Gaps	
			Corrective Actions	

Source: Safety Erudite Inc. (2019).

Figure 11.8: Detailed audit report – each element.

References

Roncea, C., (2016). Management systems audit in the annex SL context. SRAC Cert S.R.L., Bucharest. *Romania The TQM Journal*, 28(5), 786–796. ©EmeraldGroup Publishing Limited 1754–2731. Retrieved January 20, 2019 from Emeraldinsight Database. doi:10.1108/TQM-10-2015-0129.

Safety Erudite Inc. (2019). The Integrated Process Management System Provider. Retrieved June 2019 from www.safetyerudite.com

12 OEMS IMPLEMENTATION AND SUSTAINMENT SOFTWARE PLATFORM

Overview

Effective implementation and sustainment of an OEMS depends on the availability of supporting management system software to meet the business needs for business planning, asset integrity management, and safe operations. The three main components of software need of the organization therefore are as follows:

- Business stewardship – financial and business planning needs
- A computerized maintenance management system (CMMS)
- Software for managing safe operations

In this chapter, the authors explore software application challenges faced by businesses in supporting the business needs. The authors also make recommendations on software applications that may best meet business needs for organization pursuing OEMS.

12.1 Software Challenges Faced by Organizations (Clients)

Overview

Software options available to organizations comprises of the following options:

- Costly non-scalable options that are not suitable for small organizations
- Specialized systems that require clients to customize their processes to match the software provider's processes
- Access to software from remote locations
- Internal resources to support and maintain locally hosted software
- Software providers that require bolt-on options in order to provide a comprehensive solution for OEMS
- Nonintegrated modules that are not very responsive to the business needs
- Non-user-friendly processes
- Legacy software developed in-house over time for meeting the business needs

Client Responses to Challenges

Organizations and stakeholders may respond to the challenges presented above in the following ways:

- Organizations may:
  - Defer/slow down the implementation of OEMS
  - Purchase costly options that do not meet the business needs and are forced to retain the services of the provider as a means of justifying the purchase

- Business units or departments may:
  - Develop in-house, nonintegrated, labor-intensive solutions that are often people dependent
  - Underutilize the application using only pieces of it that may be helpful to improve performance in certain areas
- Users may become frustrated and lose interests in OEMS and its benefits over time

12.2 Establishing the Business Needs

Overview

While the ideal situation would be a single software platform to meet the business needs, the sad reality of the current supplier environment is no available provider of a fully integrated software platform. In view of this, organizations are generally forced to acquire multiple pieces of software to meet the needs of the business. When this occurs, it's imperative organizations that establish the business needs for software before selecting their software providers.

Defining the Business Needs

When defining the business needs, organizations should consider the following:

- How well the software meets the business needs for the following:
  - Stewardship
  - Computerized Maintenance Management System (CMMS)
  - Safe operations
- How integrated the software is in meeting business needs
- The user-friendliness of the software
- Access limitations
- Customization capabilities of service provider
- Bolt-on capabilities of service provider
- Support capabilities and responsiveness of the service provider
- Referrals from users and clients of the service provider
- Cost considerations

12.3 What to Include in Your Platform

Defining the Business Needs

A review of the current supplier environment shows no single provider of a fully integrated management system software platform. In view of this, organizations are forced to acquire multiple pieces of software to meet the needs of the business. When this occurs, it's imperative organizations that establish the business needs for software before selecting their software providers.

When establishing the business needs, the following considerations should be applied.

	Considerations
Business stewardship	• Budgeting at the equipment/unit/departmental/corporate levels • Procurements • Enterprise risk management • Audits and self-assessments • Business collaborations • Document and information management
Reliability – CMMS	• Equipment hierarchy • Defining critical assets • Preventive maintenance management • Unit cost management • Critical spares management

Safe operations	• Incident management
	  ○ Reporting
	  ○ Recording and
	  ○ Investigation
	• Sharing of knowledge
	  ○ HSE
	  ○ Best practices
	• Risk management
	  ○ Operations and environmental risk management
	  ○ Process Hazard Analysis (PHAs)
	  ○ Management of change
	  ○ Permit to work process
	• Training and competency assurance management
	• Corrective actions management and verification
	  ○ Incidents
	  ○ Shared knowledge
	  ○ Audits and self-assessments
	  ○ Risk management
	  ○ Links to CMMS
	• Inspections
	  ○ Housekeeping
	  ○ Safety
	  ○ Vehicle
	  ○ Equipment – operating, fixed, construction, and maintenance
	  ○ Seasonality – winter/summer
	  ○ Ecological
	  ○ Waste
	  ○ Equipment integrity

12.4 Selecting Your Management System Software

Test Run the Software

When selecting your MS software, organizations should ensure the software is adequately tested by users before a final decision is made regarding purchase. Complete testing is absolutely essential in order to affirm the right technology is selected for long-term application. Investment in software is a long-term investment for the following reasons:

- It is generally quite costly to acquire the software.
- It requires a lot of time and effort by the organization to implement properly.
    - Evaluation and selection
    - Programming and integration
    - Training
    - Data transfers from prior or legacy software
- Once in use and data loaded into the software, it becomes a challenging exercise to switch to alternative options.
- The avoidance of resistance by users who may view the switching process extremely difficult and debilitating when the wrong software is selected.

Supplier Support

Supplier support is required when selecting OEMS software. Dedicated software support services by the supplier are required to ensure a smooth integration of the software into the business. Support is required to ensure the following:

- Training and help is available to users of the software
- Timely response to integration glitches and failures
- Customization needs of the business
- Minimization of challenges resulting from frustrated users and associated negative messaging within informal user networks

Users versus Corporate Decision Makers

Traditionally, selection of OEMS software is led and determined by leaders within the corporate organization often with little or no input from users. In many instances too, decision makers are limited in their understanding of the needs of users. When selecting OEMS software therefore, users must be an integral part of the selection process and their voices must be heard and addressed. The needs of users should veto corporate leader's views and challenges in the selection process.

13 MANAGING AND SUSTAINING PERFORMANCE

Overview

Once implemented, OEMS requires continuous monitoring and stewardship for sustainment. Recall, as shown in Figure 1.3, that OEMS requires more than 8 years of continuous attention before it can be self-sustaining as the culture of the organization. During the formative years, however, senior leaders are required to monitor progress and implement corrective actions to continue to progress the organization toward operations excellence.

This process requires the following:

- Holding self and others accountable for performance delivered in a structured organizational team as shown in Figures 7.7 and 7.8
- Effectiveness of implementation of all requirements of each element of the management system. This is achieved through
    - Inspections
    - Self-assessments
    - 2nd - and 3rd party audits
- A disciplined approach to closing gaps that may have been overlooked in the implementation process
    - Includes resourcing and follow-up to verify closure

13.1 Stewarding OEMS Implementation and Progress

Overview

It is imperative that leaders continue to steward the progress of OEMS on an ongoing basis. Some priorities may require daily, weekly, biweekly, monthly, quarterly, and annual monitoring and progress reviews. In all cases priorities must be resourced and assigned, and leaders must be held accountable for performance and results. It is important to allow reasonable opportunities for results and performance.

Creating a Performance Dashboard

An electronic dashboard that makes it easy to monitor performance of both leading and lagging indicators related to OEMS priorities is essential for ensuring long-term success and sustainability. An online dashboard allows leaders to achieve the following:

- Real-time awareness of all parameters being monitored and measured for both leading and lagging indicators
- Easy access to stewardship information that permits timely intervention and actions
- Elevation of priorities that may not be receiving the right levels of attention and support
- Hold OEMS leaders accountable for performance on an ongoing basis

Dash Boarding

Stewardship is enhanced by having a simple dashboarding process that provides leaders with quick and easy access to trends and information relative to OEMS key performance indicators (KPIs) being stewarded. Dashboards should be able to generate real-time data and status of leading and lagging KPIs.

Dashboard Suggestions

Among the dashboard suggestions are trends for the indicators defined in the following table:

Indicators	Sample Dashboard Indicators/Trends
Leading	• Time spent in the field by leaders • Percentage of corrective actions closed within assigned periods • Number of self-assessments performed • Number of inspections completed within the period • Number of proactive interventions to prevent unsafe work • Number of identified bad actors addressed • Number of shared learnings from external sources shared across the site/Business Unit (BU) • Number of contractor safety forums held for the reporting period

Lagging	• HSE statics
	   ○ Total recordable injury frequency
	   ○ Employee recordable injury frequency
	   ○ Contractor recordable injury frequency
	   ○ Serious injuries and fatalities (SIFs) recorded
	   ○ Loss of primary containment (LOPC) frequency
	   ○ Number of incidents reported to regulators
	• Unplanned outages
	• Equipment outages
	• Number of preventive maintenance (PM) work orders in in backlog

13.2 Establishing and Managing Element KPIs

Stewardship Strategy on KPIs

Figure 13.1 provides an overview of the approaches to stewardship and management of KPIs. In this figure, the elements of the management system MS are categorized as follows:

- People
- Processes and systems
- Facilities and technology

Leading and lagging indicators are defined for each of the category and managed accordingly. Leading indicators are actively managed and stewarded, whereas lagging indicators are actively monitored.

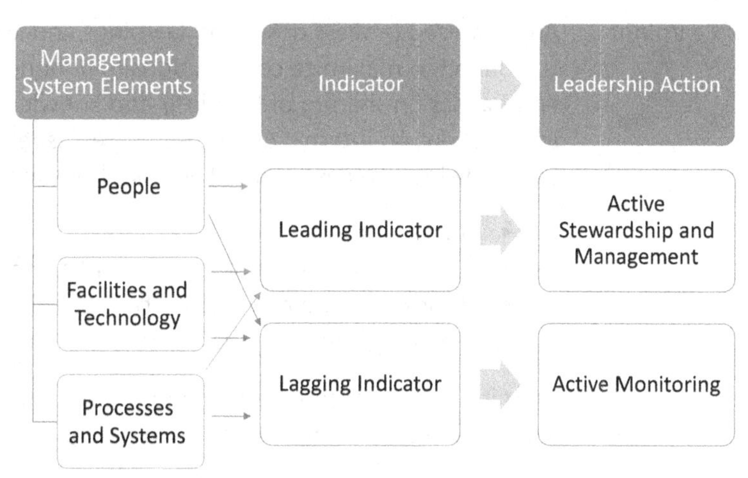

Source: Safety Erudite Inc. (2019).
Figure 13.1: Approaches to stewardship and management of KPIs.

Stretching Targets

Stretch targets are an effective means for driving continuous improvements in business performance (Thompson, Hochwarter, and Mathys, 1997). Thompson et al (1997) suggested stretch targets, "enhances motivation, performance, and creative decision making" (p. 48). They also cautioned that stretch targets do not guarantee success.

Two decades later, stretch targets continue to be a means used by organizations for continuous improvements. However, stretch targets are supported by other creative approaches to driving continuous improvements that include networks and communities of practices, transformational leadership behaviors, and advances in technology. These have led organizations along the path of continuous improvements in a sustainable way.

Donlon (2008) advised that when selecting stretch targets, internal and external benchmarks should be used to determine targets and motivating managers to achieve stretch targets.

13.3 Driving Continuous Improvements

Leveraging Networks

Networks, when activated properly, are designed initially to implement OEMS and transition to generating continuous improvements when implementation is complete. Networks should be focused on high-value, complex opportunities for developing creative solutions and improved processes. Working together, they must focus on business unit risk-based prioritized deliverables.

Generating Best Practices

For integrated organizations with multiple sites, the potential to identify highest and lowest performing units create many opportunities for identifying internal best practices that may be shared across the entire organization. Best practices are identified when silos are broken down and all business units and facilities of the asset pursue a common corporate goal, generally profit or value maximization. Once identified, internal and external best practices should be shared consistently across all business units and facilities.

Sharing of Learnings

An organized process of sharing learnings and knowledge across the organization is required to maximize contributions to continuous improvements. The emphasis is on sharing of learnings and knowledge as opposed to data and information. An internal organized process is required to ensure learning from the following sources is shared across all business units and facilities:

- Incident alerts
- Incident investigation reports and summaries
- Process improvements and best practices
- Learning from external sources

Learnings and best practices should not stop at sharing; rather, every effort should be made to implement them where applicable.

13.4 Surveying Cultural Change – OEMS and Safety Culture

Overview

Operations excellence and safety culture surveys are excellent methods used for understanding and addressing factors that may inherently adversely impact the organizations OEMS implementation and sustainability. OEMS and safety culture surveys shall consider the following:

- Structure and governance of the OEMS program
- Clarity and understanding of the strategic intent
- People – skills and competency
- Culture – values and behaviors
- Information and technology
- Process effectiveness

Frequency of Surveys

OEMS and safety culture surveys should be conducted at least once every 3 years as determined by the organizational leadership team to achieve the following:

- Differential perceptions across various levels of the organization
- Progress made relative to key priorities
- Identification of opportunities for improvement

Review Findings and Implement Corrective Actions

Findings of cultural surveys should be shared with all participants in a face-to-face forum where leaders should discuss findings with teams, highlighting successes, opportunities for improvement, and required corrective actions. Leaders should be transparent and candid about the cultural problems faced by the organization and engage the workforce in developing the right solutions and corrective actions for the issues identified. Corrective actions should be prioritized, resourced, and closed based on assigned time lines.

13.5 Linking Performance to Pay

Delivering OEMS Priorities

Once a gap analysis has been completed and work prioritized, delivering OEMS requires a considerable amount of work by those directly involved in the process. Pay for performance is an effective means of achieving the following:

- Delivering on business (OEMS) priorities
- Improving the organizational culture
- Rewarding the right behaviors.

Most organizations today maintain linked pay for performance organizational priorities. Corporate targets are established which cascade down through the business unit, facility, and eventually to the individual worker in his/her work plans. As shown in Figure 13.2, there should be full alignment between corporate OEMS priorities and the employee priorities.

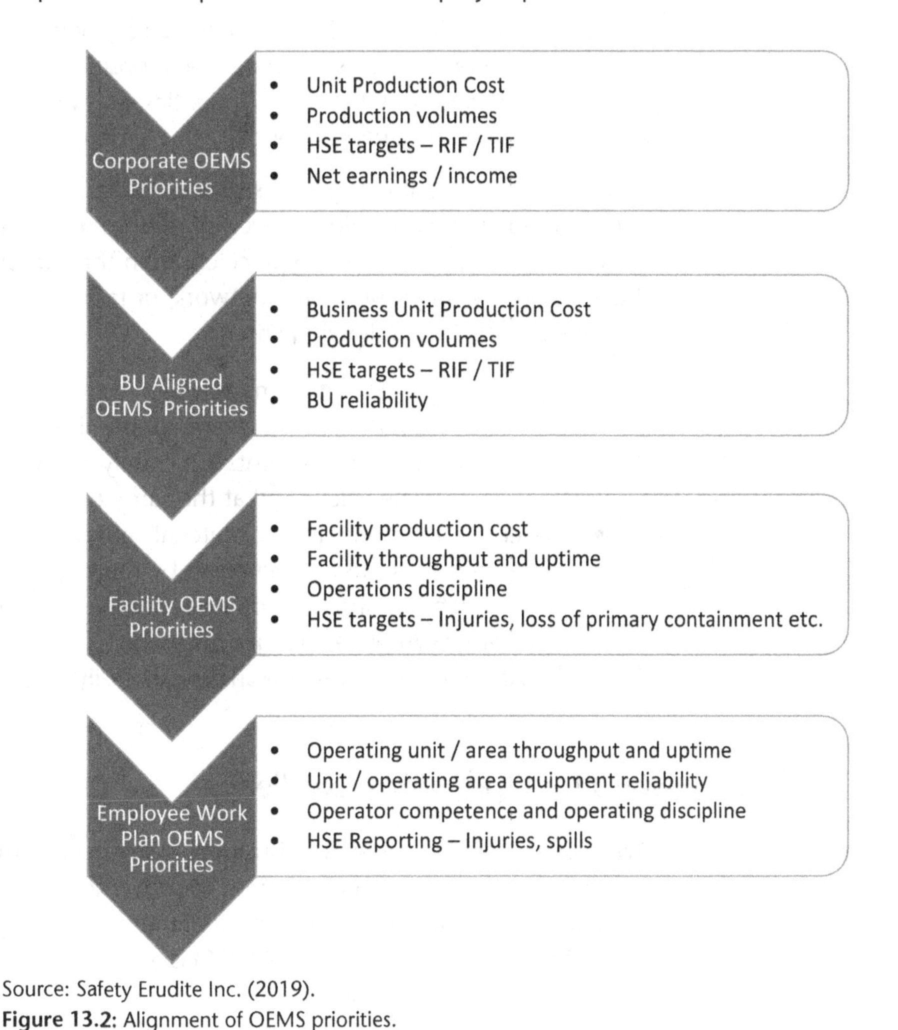

Source: Safety Erudite Inc. (2019).
Figure 13.2: Alignment of OEMS priorities.

Network Incentives

Where networks are assigned to deliver high-priority OEMS priorities across the entire organization, over and above pay for performance, financial incentives may be effective in maintaining the sustained heavy lift required for achieving success. Organizations have applied 10%–15% incentives to network members during the implementation of OEMS for successful delivery of work. Key to success in applying such incentives is the need for a robust accountability and work management process for stewardship of work so that people receive rewards for effort and performance. Failure to maintain accountability and demonstrated success creates an unacceptable financial burden and a culture of acceptable inefficiency.

Rewarding the Right Behaviors

Behavior observation programs are very useful in supporting OEMS outcomes particularly for HSE deliverables. Rewarding desired behaviors is usually an effective means of creating and sustaining desirable safety behaviors. These are often difficult to implement but highly effective in developing and sustaining the safety culture of the organization.

13.6 Celebrating the Wins

Overview

Managing and sustaining OEMS requires a disciplined approach to celebrating small wins. Celebrating wins generates some positive momentum and energizes the organization toward its OEMS goals, priorities, and vision. Workers all want to share in the success and be a part of positive achievements within the organization.

Celebration of wins also serves to showcase the business performance encouraging enablers to higher levels of performance. Celebration can vary from the elaborate, corporate level type celebration throughout the organization to the individual, workgroup, team, network, or facility celebration. Celebrating wins should meet the following criteria:

- Specific achievements or milestones
    - Celebrate one achievement at a time
    - Where achievements are closely related, multiple achievements may be celebrated at the same time
- Celebrations should reflect material achievements
- Celebrations should reflect effort by the organization versus outcomes over which the organization does not exert control – for example, market-driven price changes over which the business has no control
- Must be timely – where workers can easily recall the achievements

13.7 President's Award for Operations Excellence

Overview

Awarding excellence within the business is an effective means of driving creativity, innovation, and continuous improvements within the business. Introduced after OEMS has been implemented in the organization, President's Award for Operations Excellence (PAOE) is a very effective means of sustaining operational excellence momentum.

Criteria for evaluation and selection are established early in the year and communicated throughout the organization. Participation is encouraged and work initiatives, projects, and deliverables are aligned with OEMS priorities.

PAOE is generally offered at the following levels:

- Individual worker (employee)
- Work group, team, or department
- Site or facility
- Contractor organization

PAOE should be an annual event and should be an event. The award should be significant enough to inspire all members of the organization to seek it.

PAOE Evaluation Criteria

The PAOE evaluation criteria must be objective and fair so that all levels of the organization have an equal opportunity for winning the PAOE. Criteria for selecting PAOE recipients are provided in the following table:

PAOE Recipient	Selection Criteria

Individual worker (employee)	• Delivered exceptional individual performance
	• Nominated by peers, leader(s), and department
	• Delivered work that
	   ○ Improves work processes across the organization
	   ○ Creates a safer workplace
	   ○ Reduces operating and business risk
	   ○ Reduces cost
	   ○ Improves productivity
	• Delivered creative product or service improvements
	• Improved departmental or site/business unit or corporate business performance

Workgroup, team, or department	• Delivered exceptional workgroup, team, or departmental performance
	• Nominated by site or BU leaders
	• Delivered work that
	   ○ Improves work processes across the organization
	   ○ Creates a safer workplace
	   ○ Reduces operating and business risk
	   ○ Reduces cost
	   ○ Improves productivity
	• Delivered creative product or service improvements
	• Improved departmental or site/business unit or corporate business performance

Site or facility	• Delivered exceptional internal or externally benchmarked performance
	• Nominated by corporate leaders
	• Delivered work that showed greatest performance improvements:
	   ○ HSE performance
	   ○ Reduces operating and business risk reduction
	   ○ Reliability and integrity performance
	   ○ Delivers creative product or service improvements
	   ○ Operating cost reductions
	   ○ Operating efficiency improvements
	   ○ Best-in-class performance
	• Improves site/business unit or corporate business performance

Contractor organization	• Compared among peers, delivered on corporate aligned OEMS performance targets
	• Nominated site and BU leaders
	• Delivered work that
	  ○ Managed cost consistent with corporate OEMS expectations
	  ○ Demonstrated operations discipline behaviors
	  ○ Maintained safe and environmentally sound work practices
	• Delivered excellence in assigned work and project deliverables

Selecting PAOE Recipients

Selecting PAOE recipients must be a transparent, objective, and fair process.

- An evaluation and assessment team of trusted senior leaders adjudicate and administer the process.
- Nominees in each category are assessed objectively and compared *apples to apples* to ensure an objective process is maintained.
- Lobbying is prohibited.
- Transparency is maintained by a defensible process.
- It looks and feels right.

PAOE Award Ceremony

A PAOE award ceremony that showcases recipients of the president's awards is the culminating highlight of the process. Held at the end of the performance evaluation year, the award ceremony should be designed to achieve the following:

- Highlight and communicate exceptional performance and best practices to the entire organization.
- Create employee pride in the overall success of the organization.
- Link success to OEMS initiatives in an indisputable manner.
- Create momentum for the next year's awards.

References

Donlon, B. (2008). Setting the Right Stretch Targets. *DM Review, New York. 18*(1). Retrieved March 01, 2019 from ProQuest Database.

Thompson, K. R., Hochwarter, W. A., & Mathys, N. J. (1997). What makes them effective? *The Academy of Management Executives*, *4*(3).

Safety Erudite Inc. (2019). The Integrated Process Management System Provider. Retrieved June 2019 from www.safetyerudite.com

14 SUMMARY AND CONCLUSIONS

Overview

In this book, the authors sought to present users with a structured approach for getting OEMS implementation right the first time. The authors urge leaders to recognize that developing an OEMS requires a long-term commitment and a disciplined approach to work. The authors also recommend setting up networks, ensuring careful selection of its members to perform the heavy lifting during the implementation process. Network focus eventually transitions to continuous improvements once OEMS has taken roots within the organization.

OEMS requires a committed effort toward applying the key tenets of the plan–do–check–act (PDCA) model.

- Plan – Planning efforts are focused on clear definition of requirements and readying the organization for change.
- Do – Doing focuses on effective implementation of each element of OEMS. Networks are set up to implement OEMS and are adequately resourced and stewarded to meet prioritized deliverables. A managed pace to implementation is adopted.
- Check – Checking follows a series of approaches that includes inspections, self-assessments, second-party audits, and third-party audits. The maturity of each element of the management system is determined by comparing actual field status with requirements. Findings (deficiencies) are categorized based on PACA (procedure, accountability, competency, and assurance) principles and for targeted corrective actions. Internal and external benchmarking is practiced to determine industry standing and drive competitive advantages. A baseline management system maturity is also established that helps the organization gauge its annual progress.
- Act – Acting refers to leadership responses to deficiencies identified. Gaps (deficiencies) are risk prioritized for corrective actions and best practices leveraged across the organization.

Leaders are reminded that OEMS places a huge draw on the resources of the organization based on its baseline status. Patience and persistence are recommended since operations excellence is a journey based on timely, specific, and managed actions designed to sustainably close gaps and leverage best practices across the organization.

Regardless of the industry, organizations have been following one or more forms of management systems for decades in their pursuit of organizational success and profitability. Some have fared better than others in developing a sustainable management system. Today's competitive business environment has forced organizations to look internally and externally for ways and means of differentiating themselves from competition to maximize stakeholder value and business performance.

There is very little formally published literature on OEMS. For this reason, reference literature in several sections of this book is limited and is based on the collective experience of >100 years of the authors who worked together to assimilate and share the knowledge provided in this book to those who may find it useful. Nevertheless, OEMS is heard across the globe as organizations seek to know its details and how to successfully implement it.

Operations Discipline

Operations discipline is about doing the right thing the right way every time. This is the greatest difficulty faced by most organizations. It requires organizations to systematically document the organizational requirements packaged in 8–12 elements effectively communicating these requirements to the entire organization. Each element is assigned to a senior leader for stewardship. Documented requirements become the way for telling every worker the right thing. The right way is defined in procedures and safe work practices. Operations discipline comes from committed efforts to training and competency assurance, supervision, inspections, and a range of other business practices to ensure people are doing the right things the right way every time.

Demonstrated Success – Recent OEMS Adopters

Greg Goff, Chairman, President, CEO, Tesoro Corporation and his leadership team in 2015 pointed to successes of the organization linked to OEMS. Keith Casey, EVP Operations, Tesoro Corporation advised that sustaining performance is predicated upon its investments "in our operations excellence management systems, which is, simply stated, our game plan for how do we operate, and it provides a consistency and a base for continuous improvement, so it gets better this shift and the next shift and well into the future" (Fair Disclosure Wire, 2017a).

As mentioned earlier in this book, OEMS is a disciplined approach to improving safety and reliability performance across the business that eventually translates into excellence in business performance. These goals are echoed by George Damiris, CEO, President and Director, HollyFrontier Corporation. The CEO advised of an overarching corporate goal of being a "safe, environmentally compliant and reliable operator" (Fair Disclosure Wire, 2017b). Mr. Damaris further advised,

> we've added a lot of resources over the past several years, created technical networks. And we're implementing best practices across our refining fleet by rolling out an operations excellence management system, and it's showing really good results. After 30 years, I still love to make gasoline and diesel, and I'm really proud of my team's efforts to do this, more of it, safely, more predictably and more cheaply than we've done it in the past
>
> (Fair Disclosure Wire, 2017b)

This language is consistent with the taste of early OEMS. The impact of OEMS is reflected in teamwork, networks, safety performance improvements, predictability, and overall business performance.

Another key outcome of OEMS is a single, uniform, and consistent culture across all parts of the business. James M. Stump, SVP of refining, HollyFrontier Corporation maintains responsibilities for "refining operations, capital projects and environmental health and safety" and is an advocate of OEMS. Stump supported this outcome of OEMS when he advised "I believe we've built a very solid and strong team. We've made a lot of progress turning 6 refineries with various legacies prior to our merger into one cohesive refining complex" (Fair Disclosure Wire, 2017b).

The Importance of Patience

OEMS requires a project management approach with project life cycle period (implementation through testing and verification) of approximately 8–10 years of dedicated commitment. Patience and perseverance are therefore absolute requirements to ensure success with developing an OEMS. Minimizing cost and reaping the benefits of OEMS requires an approach that focuses on getting it right the first time. Failure and repeat of the process leads to credibility challenges and apathy among workers with an intensely more difficult process of starting over.

Learning from the Experiences of Others

Organizations venturing into OEMS should seek to learn from others who have pursued OEMS. This means doing the following:

- Review published work that summarizes the experiences of other companies.
- Hire OEMS-experienced personnel from other organizations that have been successful in implementing and sustaining OEMS:
    - Avoid hiring OEMS from different companies.
- Use consulting services with experience in OEMS who have been part of all stages of OEMS planning through implementation.
    - Unless absolutely necessary, avoid changing them during the development and implementation stages.

Managing the Organizational Change

Perhaps among the most important components of the OEMS journey is managing the organizational change associated with OEMS. Change management (organizational change) requires leaders to create and share a compelling OEMS vision with the workforce. With a shared OEMS vision, workers all aspire toward the new future of operations discipline and excellence. However, change management requires a disciplined approach by leaders to ensure all change initiatives follow a stakeholder impact assessment and mitigation of associated stakeholder risks. Once done effectively, the momentum grows toward achieving the OEMS vision and goals.

OEMS – Is It the Future?

The authors are of the view that regardless of industry, OEMS is the wave of the future and is here to stay. What this means is that organizations must adopt a fit-for-purpose variant of the approach discussed in this book for consideration and implementation to excel in today's highly competitive markets. The authors acknowledge there is no single way for implementing developing an OEMS. However, from the collective experiences of the authors and the review of existing literature on the topic, we believe that the approach proposed in this book is perhaps the most comprehensive approach for developing a sustainable operations excellence management system that is designed to deliver exceptional Health, Safety and Environment (HSE), reliability, and overall business performance.

Concluding Remarks

In conclusion, the authors are grateful for the many direct and indirect contributors who have helped shape our thoughts, knowledge, and expertise in management systems. We are grateful for the opportunity to share these thoughts with you and hope that you may take away at least one nugget of knowledge that may help improve HSE and reliability performance in your organization and everyday life.

Thank you.

References

Fair Disclosure Wire (2017a). Tesoro Corporation 2015 Analyst and Investor Day – Final. *Newspaper Articles; Linthicum*, December 09, 2015. Retrieved March 12, 2019 from the University of Phoenix Library.

Fair Disclosure Wire (2017b). HollyFrontier Corp Analyst Day – Final. *Newspaper Articles; Linthicum*, December 07, 2017. Retrieved March 12, 2019 from the University of Phoenix Library.

EPILOGUE

We hope that you found value in this highly structured and purposeful book. Achieving operational excellence is a significant challenge to all organizations regardless of size, complexity, geographical distribution of assets, and competency of its leaders and employees. It is not only the question of getting it right the first time but also the massive cost and impact of failing to get it right. In a changing economic environment of disruption, absolute need for effective optimization of processes and operational discipline is required. In the near future, implementing operations excellence in some industries may be more a question survival.

In the earlier publication on the *7 Fundamentals of an Operationally Excellent Management System*, a core aspect discussed was high-reliability organizations (HROs). HROs are organizations that can maximize value to its stakeholders within an environment of significant risks and competition. High risks can be systematically dealt with in an operationally excellent environment. However, as stated throughout the sequential chapters in this book, it is a question of a building of a system which integrates all the various aspects of an organization's workings and has people as part of the system in every single step of the way. A summary of each chapter is provided in the following table:

Chapter	Summary
Chapter 1	Chapter 1 addresses the challenges, the problem statement that faces all organizations. Building a new structure to augment a current set of systems – and, more importantly, thinking – is a daunting and challenging task. This chapter provides the discourse that organizations need to address when starting to think very seriously about the implementation of an OEMS.
Chapter 2	Chapter 2 is very important and should be read with great attention. Succinctly delivered in Figure 2.1, the model and the overall approach are summarized in our view wonderfully. The elements, standards, and the buy-in process are described in great detail. It discusses the most important aspect of OEMS, leadership, stewardship, and ownership.
Chapter 3	The "making ready" ingredients are so well set out. As one can see, the organizational culture transformation must start as soon as the organization decides to proceed with OEMS. It is not about building a system and then training leaders and employees to implement it, but about them building the system practically with the subject matter experts.
Chapter 4	Chapter 4 demonstrates the value of salvage and augmentation. All organizations operate with some degree of efficiency, but in operational excellence it is about creating a transformational change which goes beyond management systems, compliance, and governance. This is about creating the structure and system which brings all the great systems in place together, aligning them with achieving a generative (safety) excellence culture where stewards usher the organization to a new level of operating integrity.
Chapter 5	When we write systems, policies, procedures, and operating instructions (or any other document) we must start with the end in mind; the end being the end user and not some certificate of conformance! Sophisticated documentation may impress auditors but, like a sword, it is double-edged: it neither makes it easy to implement nor translates well into action. Writing the systems and procedures in an effective way that ensures effective implementation and improves performance is what all organizations must aspire to do.
Chapter 6	We must make sure we have primed the organization to receive, accept, believe, and implement OEMS. This chapter explains the preparation required to ensure success in this major undertaking and talks about how to prepare the organization for that task. It leverages and makes practical many of the well-grounded change management models and builds on the management and social sciences associated with effective change management. This chapter is important for so many in the organization because OEMS is not a management system, it is cultural transformation.

Chapter 7	Chapter 7 discusses establishing of networks and communities of practice – these are among the most fundamental cornerstones of OEMS. It not only talks of networks and network structures but also provides a guide to the roles that must be played by the organization's actors – its leaders, employees, and the champions of the transformational change.
Chapter 8	Preparation is one of the most significant success factors in OEMS. Many organizations decide to proceed and lack the patience at times to plan, plan, and plan. Preparation is fundamental to the priming process, but also to the setting of the architecture, both the management systems and organizational (people) infrastructure. Leadership must appreciate that the pre-planning and preparation processes must be given due support.
Chapter 9	Chapter 9 provides extremely valuable guidance on how to use all this very valuable data that the organization already has – it explains the gap analysis process and how to use it to assure success. It is about leveraging on the big wins, helping an organization make the best of what it can do in the shortest period.
Chapter 10	Here, the authors provide a systematic approach to the implementation of OEMS in a measured way. The approach requires setting the organization up for success by identifying and fixing the must-haves and known gaps identified by the business to sustain current levels of operation and business in a safe and reliable manner.
Chapters 11 and 12	Chapters 11 and 12 provide assurance through compliance and how to undertake this effectively. The use of software and data management systems is also discussed as this has become fundamental to all organizations; we continue to rely more and more on information technology and management systems technology.
Chapter 13	Chapter 13 discusses how to sustain OEMS as implemented. It encourages organizations to steward performance and to celebrate successes and wins. It encourages organizations to recognize those who are involved in driving the organization to operations discipline and excellence.
Chapter 14	This book has a unique value proposition. It has been written by highly experienced, involved, and passionate practitioners. The book is full of succinct information supported effectively with operational excellence application experience from so many successful organizations for the past 20 years. The information has been written for users. There is a paucity of published materials on operational excellence, and there are none which are as easy to follow and use to implement as this. This book transcends theoretical concepts; it is about the practical application of operational excellence practice and knowledge. A management system is only as good as its ability to produce the value proposition it was designed to deliver.

Ahmed Khalil

Director, Fire and Safety, Bahrain Petroleum Company (BAPCO)

INDEX

G

H

I

K

L

M

N

O

Printed in the United States
by Baker & Taylor Publisher Services